No. 1685
$17.95

# THE
# COMPLETE GUIDE TO
# SATELLITE TV

### BY MARTIN CLIFFORD
### TECHNICAL CONSULTANT
### KLM ELECTRONICS, INC.

**TAB BOOKS Inc.**
BLUE RIDGE SUMMIT, PA. 17214

FIRST EDITION

FIRST PRINTING

Copyright © 1984 by Martin Clifford

Printed in the United States of America

Reproduction or publication of the content in any manner, without express permission of the publisher, is prohibited. No liability is assumed with respect to the use of the information herein.

Library of Congress Cataloging in Publication Data

Clifford, Martin, 1910-
   The complete guide to satellite TV.

   Includes index.
   1. Direct broadcast satellite television.  2. Home
video systems.  3. Television—Receivers and reception.
I. Title.
TK6677.C58  1984    621.388'5    83-24270
ISBN 0-8306-0685-8
ISBN 0-8306-1685-3 (pbk.)

Cover photo courtesy of KLM Electronics, Inc.

# Contents

# Acknowledgments

Companies engaged in the manufacture and distribution of satellite TV products are aware of the need for supplying as much information about this subject as possible. Upon request they supplied substantial amounts of data, and so my thanks to: Amplica, Inc; Automation Techniques, Inc.; Avantek, Inc.; Blonder-Tongue Laboratories, Inc.; Crystal Technology; DM Electronics (Div. Design Homes, Inc. — Franklin A. Week, President); R. L. Drake Co.; DX Antenna Co., Ltd.; Earth Terminals; Echosphere Corp.; H & R Communications Inc. (Subsidiary Craig Corp); Hero Communications (Div. Behar Enterprises, Inc.); Industrial Scientific, Inc. (Jackilene Dietrich, VP); Intersat; Kaul-Tronics, Inc.; La Grange Satellite Television Systems (Jerry Brandt); Lewis Electronics Co.; Lindsay America (Consumer Electronics Div.); Lowrance Electronics, Inc.; Marshall Electronics, Inc. (Leonard Marshall, VP); McCullough Satellite Equipment Inc.; Micro-Tenna Associates, Inc; National Microtech, Inc.; Northwest Satlabs; Paradigm Mfg. Inc. (Max Schreiber, Mgr. Wholesale Mktg.); Pico Satellite, Inc. (Elmer E. Pegram, VP); Precision Satellite Systems; Sat-Pak (Steve Sheppherd); Satellite Television Corp.; Tele-Star Satellite Communications, Inc.; Total Television, Inc. (Gordon Crawght, VP, Marketing/Sales); Transfer, Inc.; Winegard Co.

# Special Acknowledgments

This book started with a casual conversation I had with Peter Dalton, President of KLM Electronics, Inc. and it was with his encouragement that this project got rolling. As the book moved along and started to take shape there were others who also made contributions. These include Gary Gordon, production coordinator of KLM Electronics, Inc.; Courtney Newton, Daniel Roher Co.; Don Berg, VP, Channel Master, Inc., (Div. of Avnet); and Larry Steckler, publisher of Radio-Electronics.

A special note of thanks should also go to M/A Com Video Satellite, Inc. They supplied the material that appears in the appendix, and which was extracted from their publication, the Television Receive-Only (TVRO) Planner, an excellent and logical presentation of a step-by-step mathematical analysis of satellite reception design.

# Introduction

At one time television meant receiving a broadcast signal. Alternate program sources weren't available. Television broadcasting started with a dozen VHF channels and subsequently expanded to include UHF, but the method of reaching the in-home TV set remained the same.

Things have changed. You can now put a picture on your TV screen with the help of a videocassette recorder or through the use of video discs. Television broadcasting is also receiving some competition in the form of "wired TV," better known as cable.

Of all of these, the one that is the most exciting and that offers tremendous potential for TV viewers is satellite TV. Satellite TV is capable of delivering far more programs than broadcast TV. Its pictures are superior since signals bouncing from buildings and terrain, resulting in "ghosts" or multiple images, are no affliction. Satellite TV is capable of supplying TV broadcast studio picture quality. Further, because of the wide band of frequencies available, satellite video, unlike VHF and UHF, is FM, supplying the noise elimination advantages so long enjoyed by FM radio. When high-definition TV finally becomes available, it will be via satellite.

It is now also possible to pick up stereo sound from satellites, and that sound can be used to drive an in-home hi/fi system. In this way we have the joining of two of our best forms of home entertainment — video and audio. By using satellite audio in conjunction with a hi/fi system, it is possible to avoid the extremely poor sound supplied by most television receivers.

Installing a satellite television receiving system does involve more work and a greater cost than the usual TV setup for VHF and UHF broadcast TV, but the results are well worth it.

This book will make you a more knowledgeable consumer and will remove the mystery that surrounds this subject. For those readers who have a technical background and who have a working knowledge of what is really elementary mathematics, there is an appendix, with design information for a television receive-only system. Finally, for those interested in the vocational opportunities of satellite TV, the material in this book supplies a starting base.

# Chapter 1

# An Introduction to Satellites

In the years that preceded Christopher Columbus, it was commonly accepted that the earth was flat. When that concept was disproved, the earth was regarded as a perfect sphere. A rubber ball is a sphere, that is, every point on its surface is equidistant from its center. The earth isn't a sphere since it isn't perfectly round. Instead, it is an oblate spheroid; that is, it simply resembles a sphere.

You can get a better conception of the shape of the earth by putting an orange on a smooth surface and then pushing down on the top surface of the fruit. The top and the bottom will tend to flatten; the area around the middle will bulge somewhat. We live on an earth that is also somewhat flattened in the areas of the true North and South Poles, with some expansion around the equator.

## THE POLES

If you could draw a line from the center point of the top part of the earth to the center point of the bottom, the line would be almost 8,000 miles long. A more accurate measurement would be 7,899 miles. This imaginary line, as indicated in Fig. 1-1, is the earth's axis. One end of this axis is the true North Pole; the other end is the true South Pole.

We can draw another imaginary line, this time from one point on the equator, through the center of the earth, out toward the other side, touching the equator again. This line would be 7,925.6 miles long. If you subtract the length of the north-south axis from the equatorial diameter the difference is 26.6 miles. It means the earth, at its center, has a bulge and this bulge is a total of 26.6 miles.

We can divide the earth into two equal parts, or hemispheres, by

Fig. 1-1. The true North Pole is the vertical axis of the earth. The equator forms a right angle with this axis and is zero degrees latitude.

drawing an imaginary line around this maximum circumference. This line, called the equator, is 24,899 miles long.

## THE EARTH AS A SIGNAL PATH

In the early days of radio, circa 1919, broadcasting was done with radio waves having frequencies similar to those used today. Such waves had, and still have, an unusual characteristic, for they tend to follow the curvature of the earth. In their travels, these waves encounter all kinds of man-made and natural obstacles, and so their energy gradually dissipates, changing from electrical to heat energy. The amount of heat involved is imperceptible.

Because of the ability of radio waves to travel great distances, newspapers, in the early days of radio, commonly printed claims for dx (distance) reception. In a few instances these were as much as 3,000 miles.

The concept of free in-home entertainment gripped the public's fancy, and while manufactured radio receivers were available, there was a large build-it-yourself movement. Radio tubes were expensive, often costing as much as $15, then considered a good weekly salary. One alternative was the crystal set, using a tuned coil, a galena detector (iron pyrites) and a single headphone. These were the first solid-state receivers, and preceded transistors by almost 30 years.

In the meantime, prior to and subsequent to World War 1, experimenters called radio amateurs or "hams" were working with radio waves having shorter wavelengths than those used by radio broadcasting. In so doing, they made a remarkable discovery.

Unlike broadcasting stations having operating powers measured in thousands of watts, lack of money restricted these amateurs to very modest amounts of transmitting power, quite often no more than 10 watts. They soon learned that even with such small power ratings they could communicate over tremendous distances, actually world wide, thus making the distance achievements of standard broadcasting stations puny by comparison. As these amateurs, starting with 200 meter waves, began experimenting with still shorter wavelengths, they found long-distance communications becoming easier and even more reliable. Ultimately, the radio amateur fraternity began operating on wavelengths from 160 meters down to as low as 6 meters.

In the meantime, radio broadcast stations, still operating on the same frequencies, realized they were losing the distance race. Increasing operating power to 5, 10 and more kilowatts seemed to make very little difference to a distance conscious public. A number of intriguing proposals were made and some were actually tried.

Since the curvature of the earth, and various obstructions on it, hindered the passage of radio waves, the obvious solution was to get as high above them as possible. Apparently, it had not occurred to operators of broadcast stations to question why radio amateurs, with much lower operating powers, were able to achieve world-wide communications. One proposal was made to use an airplane circling at a height of some 10,000 feet and to equip it with radio broadcasting equipment. Apparently, someone forgot to question how long an airplane could remain at that altitude without refueling. An alternative was a fleet of planes, and then finally a dirigible was suggested. These methods were financially and technically impractical, and so broadcast stations finally abandoned the distance quest, particularly since the nation was beginning to be covered with a network of broadcast radio.

Radio amateurs were able to achieve world-wide communications because of the behavior of radio waves, and also because of a number of ionized layers surrounding the earth, acting as a huge signal reflector.

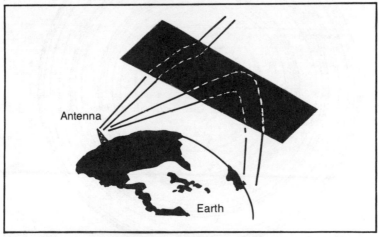

Fig. 1-2. The effect of the ionosphere on waves depends on wave frequency. For lower transmission frequencies the radio waves are refracted and bent back towards earth. As frequencies become higher the waves are more slightly refracted but can escape to outer space. For the microwave frequencies used in satellite TV the ionosphere has no effect.

At the shorter wavelengths used by radio amateurs, radio waves are no longer completely earthbound and so part of their signal energy travels out into space (Fig. 1-2). These waves, like light, can travel on out into space indefinitely, but, surrounding the earth are a number of ionized layers, ranging from 60 to 200 miles above the earth's surface (Fig. 1-3).

These ionized layers act as a boundary for radio waves, reflecting them to earth, not to their point of origin, but some considerable distance from it, a phenomenon known as skip distance (Fig. 1-4). It was this phenomenon that enabled amateurs to communicate over long distances. The problem with such long-distance communications was its lack of consistency. Further, as the wavelengths were made even shorter, radio waves were able to pass right through the ionized layers.

## CATV

Almost 30 years after the establishment of radio broadcasting, television made its appearance. Its operating range was quite limited, with 50 to 75 miles as a practical limit. For those who lived beyond these limits or who were surrounded by hills or mountains, television was an impossible dream.

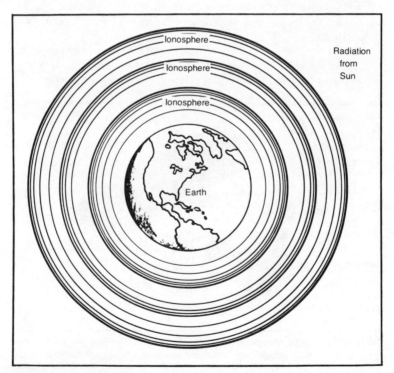

Fig. 1-3. The ionosphere contains belts of ionized layers caused by radiation from the sun.

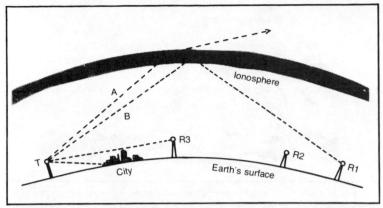

Fig. 1-4. Wave A passes through the ionosphere, although it becomes partially bent. Wave B is reflected by the ionosphere to receiver R1. The receiver at R2 doesn't pick up the signal because of interference by the earth's curvature. Receiver R3 receives the signal because it is *line of sight* with respect to the transmitter T. Receivers located in the city also pick up the transmitted signals. The distance between the transmitter, T, and the receiver at R1, is called skip distance.

It soon became fairly obvious that television waves were quasi optical, that is, behaved somewhat like light waves. The TV waves bounced off high buildings and high terrain. In short, what was necessary for decent TV reception was a clear line of sight between the television transmitting antenna and the television receiving antenna.

The solution to the TV problem for those living in fringe areas was to have a master antenna installed as high as possible, preferably on top of a hill or mountain. The signal was then routed to subscribers' homes via coaxial cable, with one or more line amplifiers used to make up for signal losses produced by the cable (Fig. 1-5).

This was a first step away from free TV, but necessary, because of the high cost of installation of the master antenna, the connecting coaxial cable, the amplifiers, and the expense of wiring the cable into viewers' homes. Known as community antenna television and abbreviated as CATV it had limited success, since it wasn't practical in all instances.

CATV did represent one important change. It was the first move away from a totally broadcast system to one that was a combination of broadcast and wired reception. Aside from the high installation cost, CATV systems were totally dependent on the transmissions of just a few television stations, possibly just one or two.

## THE CLARKE PROPOSAL

In 1945, a noted science fiction writer, Arthur C. Clarke, aware of the communications limitations as they existed at that time, and having a knowledge of the reflecting layers of ions surrounding the earth, proposed a

5

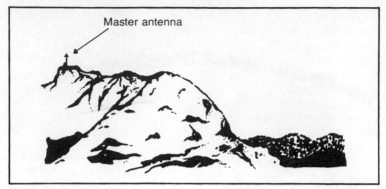

Fig. 1-5. In a CATV system the master antenna is mounted as high as possible. The television signal is brought to the homes of subscribers via coaxial cable.

satellite communications system in the October 1945 issue of a British technical magazine, *Wireless World*. His suggestion was to place three reflectors in space such that signals striking them would be reflected to almost any area on earth. These devices were to be situated 22,300 miles out in space, positioned above the equator.

The selection of that distance wasn't a haphazard guess. The reflectors weren't intended to be fixed in space, but in orbit around the earth such that their orbital speed would correspond to the rotation of the earth. The effect would be that of reflectors hovering in space, as though fixed in position. Today, the circular path followed by satellites is known as the Clarke orbit.

## WHAT IS A SATELLITE?

If you want to see a satellite take a look up in the sky at the moon. Don't dismiss it out of hand because it is so customary a sight, for some day it may be one of the biggest signal receiving and transmitting satellites we have.

The moon is a satellite and is our closest neighbor in space, 238,857 miles away. The other planets have satellites also: Mars has two, Jupiter has 12, Saturn has 9, and so on. Nature has supplied us with plenty of examples. These satellites are categorized as natural; man-made satellites are called artificial.

## ENTER SPUTNIK

The year was 1957, an ordinary year, just like any other. There was one event that seemed newsworthy and it was headlined in a few papers for a day, and then the world went about its business, not realizing an important step had been taken that would affect everyone's life. October 4, 1957 was the day on which the first man-made satellite was launched. Known as Sputnik, it was lofted into space by the USSR using rocket propulsion. Not to be outdone, the following year the U.S. Air Force built and launched Score, the first North American satellite.

Compared to our present-day satellites, Sputnik was unsophisticated. It was a sphere about 23 inches (58 cm) in diameter and weighed approximately 183 lbs (83.6 kg). Its orbital velocity was approximately 26,200 ft/sec(8 km/sec) making one complete revolution about the earth every 96 minutes. Its initial altitude was about 560 miles above the earth and its orbit was inclined about 65 degrees to the plane of the equator.

Sputnik was more than just an empty shell, for it contained a transmitter that broadcast signals at frequencies of 20.005 MHz and 40.002 MHz. MHz is an abbreviation for megahertz, with each megahertz representing one million cycles per second.

## SPUTNIK II

Sputnik I was followed by Sputnik II, a cylindrical object weighing about 1,200 lbs. Its orbit was highly eccentric, measuring 1056 miles at its apogee, its most distant point from the earth, and about 100 miles at its perigee, its nearest point to the earth. The satellite was equipped with a short wave transmitter and a living passenger, a dog. The dog survived for several days, with the sound of its heartbeat being transmitted.

## MOUSE

Another early, experimental satellite, was Mouse (1958), an acronym for Minimum Orbital Unmanned Satellite of the Earth. The orbital plane of this satellite was such that it spent most of its time in sunlight. This satellite had a horizontal axis of symmetry about which it revolved. It was made to point toward the sun by putting the satellite into gimbals and spinning it prior to takeoff with the axis pointing toward the sun. The purpose of this arrangement was to maximize the use of solar power since the satellite contained a solar charged battery, an arrangement that permitted charging the battery indefinitely. The use of sunlight as a source of energy was a concept that was carried forward and is used today.

The orbit of this satellite was such that it passed over the North Pole and also the South Pole. The satellite was equipped to run a number of scientific tests, storing this data during most of its orbit, and then telemetering it when near one of the poles. A receiving station located near one of the poles was able to contact the satellite, turn on its transmitter and receive the stored information. The orbital path of this satellite ranged from 200 miles at perigee to 600 miles at apogee.

## TELSTAR

The concept of transmitting a television picture from a satellite, subsequently also known as a bird, came in 1962 when AT & T launched Telstar (Fig. 1-6), the first satellite to relay a television picture. However, the concept of geosynchronous orbit wasn't put into practice until the following year when the Hughes Aircraft Corporation launched Syncom.

1957 - Sputnik, first manmade satellite launched by the USSR.
1958 - Score, first American satellite, built and launched by the U.S. Air Force.
1962 - AT & T satellite Telstar, first bird to relay a television picture.
1963 - Hughes Aircraft Corporation launches world's first geosynchronous satellite, Syncom.
1965 - Intelsat network formed to provide intercontinental data, voice and television linkage.
1966 - Early Bird (Intelsat 1) launched, capable, of 240 phone calls and 1 transatlantic TV signal.
1967 - Intelsat II. Satellite in place over Atlantic and Pacific.
1969 - Intelsat III. Capable of handling 1500 telephone circuits plus video.
1971 - 1975 - Intelsat IV series. 3750 telephone circuit capability plus 2 TV channels.
1975 - Aryabhata. Indian made satellite launched from and by the Soviet Union.
1976 - Intelsat IV-A.
1980 - Intelsat V. Capable of operating in the C and K bands.
1980 - Cosmos 1151. Ocean monitoring satellite launched by the Soviet Union.
1980 - Cosmos 1154. Monitoring satellite launched by the Soviet Union.
1983 - Insat 1-B. Communications, television, and meteorology satellite for India (Indian National Satellite System).
1983 - Galaxy 19, Telstar 3A, Galaxy 11, Sakura A2 (Japan), Sakura B2 (Japan), Palapa B1 (Indonesia), ECS-1, Telecom 1, Westar VI.

Fig. 1-6. A brief summary of important events in satellite history.

## GEOSYNCHRONOUS ORBIT

One of the elementary requirements for the continuous reception of a signal is that the distance between the transmitter and the receiver should be relatively fixed. If you are in a car listening to an FM station, for example, the greater the separation between your car's antenna and the broadcast station, the worse the signal-to-noise ratio, until finally the noise level becomes so high the signal is no longer listenable.

On the planet earth, you are in a comparable position to that of being in a moving car except that the earth's rotational speed is so much greater. The velocity of the earth in its orbit in space is about 19 miles a second. The earth also has a spin, revolving around an imaginary axis. This rotation of the earth is its diurnal or daily motion. Each complete rotation takes 23 hours, 56 minutes and 4 seconds, or approximately 24 hours (one day and one night). As far as satellites are concerned, it is this rotation that concerns us, and not the velocity of the earth in its orbit in space. That orbit, incidentally, is an ellipse around the sun and is a path of 577,000,000 miles, a movement or revolution of the earth called its annual motion. It takes the earth about 365 days and six hours or one year to complete a single, elliptical path. Naturally, we aren't aware of the speed of the earth in its orbit since we have no reference, nor is there any vibrational reminder as there is in a car or train.

To be able to receive television pictures from a satellite, its rotational speed must be such that it appears to be stationary, even though it is following a circular path or orbit.

This situation is somewhat like that of a wheel. A mark inscribed on the hub rotates and so does a smear mark on the rim. Both are turning, but the position of the smear mark on the rim with reference to the mark made on the hub remains constant. The earth is comparable to the hub; the satellite to the smear mark on the rim. Their turning rate is such that the satellite is fixed in position with reference to an area on earth. Such a satellite is known as geostationary or geosynchronous, so called because it is stationary with respect to the earth or synchronous with its orbit. The word geostationary is a compiled word based on *geo* meaning earth and *stationary* meaning fixed in position. Hence, geostationary means fixed in position with reference to earth.

The distance above the earth a satellite must reach so as to be able to travel at synchronous speed is 22,279 statute miles above the equator. (A statute mile, also known as an English mile has a length of 8 furlongs, or 1760 yards or 5,280 feet.) This is, indeed, a specific distance and applies to all telecommunications satellites. The higher an object is in space the longer it takes to travel around the earth. The moon, the natural satellite of the earth, has a perigee of 221,463 miles while its apogee distance is 252,710 miles, and requires about 28 days for a complete circumnavigation of the earth. An object in space, closer to the earth, possibly a few hundred miles, may go around the globe in just a few hours.

Between these two extremes we have a distance at which the satellite will remain fixed with reference to a selected earth area (Fig. 1-7). This doesn't mean the satellite rotates at the same speed as the earth. Since its circular orbit is so much greater, its linear speed is somewhat greater than that of the earth's surface.

## INTELSAT

The world's first geosynchronous satellite, Syncom, demonstrated the practicality of utilizing a satellite for television entertain-

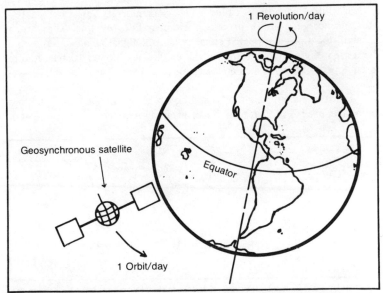

Fig. 1-7. One satellite orbital day is equivalent to one earth revolution day.

9

ment. Its height above the earth, measured in thousands of miles, far exceeded the 10,000 feet envisioned by early pioneers of widely distributed signals.

In 1965 nineteen countries, foreseeing the possibility of using a group of satellites for worldwide entertainment, formed a group called The Third International Television Satellite Organization, a title that lent itself nicely to the formation of the acronym, Intelsat. While Intelsat looked forward to supplying television, it was also interested in telephone and data communications, on an international basis (Fig. 1-8). More than 100 countries are now members of Intelsat, which is equipped with five satellites and has more than 250 ground stations. These satellites are fifth generation and so are identified by the Roman numeral V.

Currently, there are about 16 television oriented U.S. and Canadian satellites in orbit above the equator, but the trend is toward increasing their number. The Federal Communications Commission has authorized the launching of 20 more satellites by 1986. This does not mean 20 new television channels will be established since a typical satellite often has a 12 or 24 channel capability, but it could have even more.

The specific position of a satellite in space is sometimes called a satellite orbital slot, more often briefly referred to as a slot. All of the domestic television satellites, actually television relay stations, are positioned in an orbit right above the earth's equator between 79 degrees west longitude and 143 degrees west longitude (Fig. 1-9).

## ADVANTAGES OF SATELLITES

Satellites, sometimes called CATV or community antenna TV satellites, do not generate or produce their own TV programs. They aren't passive, for a passive device simply passes a signal along, often with a decrease in signal strength and without modifying it. Instead, each satellite is equipped with receivers and transmitters, hence they are active systems.

The tremendous height of a satellite's orbit and the fact that satellites are active, has resulted in a number of advantages, some of which are yet to

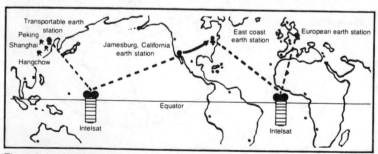

Fig. 1-8. Intelsat IV made a world-wide telecast of President Nixon's visit to the People's Republic of China in 1972.

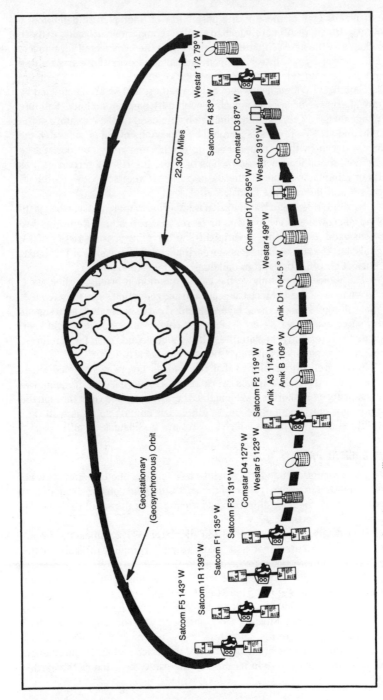

Fig. 1-9. Location of the North American satellites.

11

be realized. One of these is the possibility of finer picture definition, a potential that may ultimately lead to pictures comparable in quality to those produced by the finest color film. Even without these expected advantages, pictures sent by satellite have many advantages over those transmitted from earthbound television broadcast stations.

Satellite TV signals are above the weather, and so aren't affected by atmospherics. Nor are the signals troubled by sunspot activity. Satellite TV signals have overcome the limitations imposed by the curvature of the earth, and so TV viewers, separated by thousands of miles can watch the same program without dependency on earthbound relay stations. Large buildings, metallic structures such as bridges, interfering terrain such as hills or mountains—none of these can come between the viewer's television set and the satellite signal.

Satellite television transmission is superior to that produced by earthbound microwave relay stations or terrestrial repeaters. Repeaters are devices used to overcome the straight-line traveling characteristics of TV signals, but the fact remains that any earthbound transmission of TV signals is subject to reflections from buildings and high terrain.

The ionized layers above the earth, mentioned briefly earlier, gave radio amateurs long distance transmission and reception since they acted as signal reflectors. Since these layers shielded earth generated radio signals from outer space, wasn't it possible that these same shields would keep signals from reaching the satellites? If the signals did reach the satellites, wouldn't that shield keep them from coming back to earth?

Fortunately, it developed that the ionized layers were broadly frequency selective. As signals were made to increase in frequency—that is, using ever shorter wavelengths, the waves zipped right through the barrier, not only going up to the satellites but coming down as well. For satellite signals, then, the ionized layers are no obstacle at all.

## THE ORBITAL PATH

The amount of time for a satellite to complete one full orbit, that is, one complete circumnavigation of the earth, depends on the shape of that orbit, whether round or elliptical, and the height of the satellite above the surface of the earth.

If a satellite has an altitude of 180 kilometers (112.5 miles) its revolution time is 88 minutes, that is, it makes a single, complete orbit around the earth in little more than one hour.

At such a low altitude, the satellite encounters friction from air molecules, gradually slowing the satellite and shortening its useful working life. The satellite will gradually orbit at a decreasing height above the surface of the earth. Finally, the high speed plus increasing molecular friction will incinerate the satellite.

To produce a geostationary satellite, there are various requirements. The satellite must rotate in its orbit in the same direction as the earth's rotation, and its orbital altitude must be 22,300 miles so that its revolution time is the same as that of the earth.

## EARTH DAY VS ORBITAL DAY

Not just one satellite, but a number of them, travel around the earth, following an imaginary space line of latitude that is parallel to the equator. Just as it takes the earth a day to achieve one complete revolution, so too does it take the satellite a day to complete a revolution. One is known as an earth day; the other as an orbital day. Both require the same amount of time.

Since the diameter of the earth at the equator is approximately 8,000 miles, the altitude of a geosynchronous satellite is almost three times as much. Because of its height, a single satellite can *see* about one-half of the surface of the earth. It would take only two satellites to be able to see almost the entire surface. With three satellites positioned 120 degrees from each other and placed in an equatorial plane, the entire surface of the earth could be covered by signals transmitted from them, with the exception of a small area around the North and South Poles, to 8 degrees latitude. The significance of this is that worldwide communications can be had from any point on the inhabited globe to any other point, via satellite.

If satellites are used in a lower orbit than one that is geostationary, such satellites will not be fixed with reference to a selected area on earth. If they could be seen, they would appear to rise over the eastern horizon, move through the sky, and then disappear over the western horizon. Because of their lower altitude, more of them would be needed, since each would not *see* as much of the earth's surface.

In the early days of satellite development, lower orbits for satellites represented a practical approach since launch technology could not achieve today's geosynchronous orbit altitude.

## TYPES OF ORBITS

The orbit of a satellite around the earth is circular, since the earth makes a circular rotation on its axis. This doesn't mean that every object in space follows such an orbit.

When a satellite is first launched into space, it follows an elliptical orbit, a path such as that shown in Fig. 1-10. It is subsequently pushed into a circular orbit by thrusters on the satellite. Not only must the satellite be put into a circular orbit, but it must be edged into its proper slot, that is, it must be correctly positioned with respect to the other satellites.

Man-made satellites use circular orbits, but natural objects in space follow elliptical orbits. While the rotation of the earth is circular, its orbit around the sun is elliptical. Not only the earth, but all the other planets revolve around the sun in elliptical orbits. The moon, the earth's natural satellite, revolves around the earth, also in an elliptical orbit.

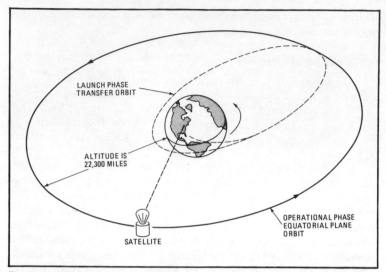

Fig. 1-10. The initial orbit of a satellite is an elliptical path. Using hydrazine thrusters, the satellite is pushed into an orbit 22,300 miles above the equator. (Radio-Electronics)

## NON-GEOSYNCHRONOUS ORBITS

Space vehicles, non-satellite devices containing test instruments only or test instruments plus one or more persons, orbit far closer to the earth than satellites. A representative orbit could be one having an apogee of 195 miles and a perigee of 184 miles.

Since the space vehicle is fairly high, the force of gravity is considerably reduced. As a result, it has become practical to deploy satellites from the space vehicle. These are ejected and self power their way into geosynchronous orbit by on-board thrusters and steering jets. These, however, are controlled from earth guidance stations. The advantage is the reduced cost of getting a satellite into orbit, plus the fact that space vehicles have the cargo room for carrying more than one satellite.

While a satellite is basically a communications vehicle, it is equipped to have some rocket properties. The satellite, however, must first be given a *lift* into space, but once it is in space it can make use of solid fuel rockets to achieve orbit and to make any necessary position corrections thereafter.

Components on satellites cannot be repaired in space, at least not at the present time. This means everything in or on the satellite must have a high order of reliability. In effect, all components must be *space qualified* and must work in microgravity conditions indefinitely, with an estimated functioning life of about 10 years.

Because of the high reliability requirements and the expense involved in getting a satellite into space, the overall cost of a satellite is extremely high and is counted in terms of millions of dollars. To this must be added the fact that not all launchings are successful, and even if they are, the satellite may not get into its proper orbit.

While satellites in space currently cannot be directly serviced and are completely dependent on correction signals sent from earth, it is possible that in the future ailing and aged satellites may be serviced on the spot by astronauts but, more likely, will be picked up by a space shuttle and brought back to earth for repair and possibly for retrofitting as well. While these things may be done and may be technically feasible, the question is always one of economic justification. The question is also one of space availability. Because of space limitations, it may become compulsory to remove a satellite that is no longer functioning or that could be replaced by one capable of an increased number of services.

## SLOT UTILITY

It would seem, offhand, that since space seems to be unlimited, that there would be an infinite number of slots available. While satellites are indeed out in space, as indicated earlier, they must be at a specific distance to be geostationary. They are presently spaced with a separation of 4 degrees, except that a pair of Canadian satellites use 4.5 degree spacing. Since there are 360 degrees in a circle, the earth could be surrounded by a total of 90 satellites.

Unfortunately, a large part of the slots made available by a complete circle are useless. If all were used, the signals of many of these would end up in oceans or in countries not equipped for satellite signal reception. Currently, the geocentric arc now being used by North American satellites is less than 90 degrees, and the number of available slots is at a premium.

## SEPARATION OF SATELLITES

The satellites are arranged in a circular orbit in the equatorial plane. This plane is an imaginary geometric two-dimensional figure that slices the earth in half perpendicularly along the line of the equator.

It is easy enough to calculate not only the separation between each satellite, in miles, but the orbital speed of the satellites as well.

At the equator, the diameter of the earth is 7,926 miles, and half of this, the earth's radius, is 3,963 miles. The distance to a satellite is 22,279 statute miles. Adding the radius of the earth at the equator to the distance of the satellite from the earth's surface, we get 26,242 miles. Based on the formula that the circumference of any circle is equal to $2 \times 3.1416 \times$ the radius, the orbital circumference is then equal to $6.2832 \times 26,242$ miles $= 164,884$ miles. This last number is the distance traveled by each satellite in its orbit every day.

Presently, most North American satellites are spaced at an angular separation of 4 degrees. Since the orbital path of the satellites is a circle and since every circle, regardless of size is a total of 360 degrees, the distance from one satellite to the next is $4/360 \times 164,884 = 1,832$ miles. There is currently a proposal to use an approximately two-degree separation in an effort to increase slot space availability. Under these circumstances, each satellite would be 919 miles apart.

The path followed by each satellite isn't smooth and undeviating. There may be a satellite wobble of as much as 60 miles, but this is insignificant compared to the traveling distance of the satellites.

As indicated earlier, the orbital day of the satellites is the same as a terrestrial day and is 24 hours long. The circumferential speed of each satellite is $164,884/24 = 6,870$ miles per hour. The circumferential speed relative to earth, though, may be considered zero.

## LINE OF SIGHT

The number of satellites anyone can *see* is limited to line of sight. Consequently, it isn't possible to pick up the signals of a group of satellites, for example, known as the Intelsat Indian Cluster, positioned between 50 degrees and 70 degrees east latitude. However, it would be possible to pick up their signals if they should transmit them to satellites we can *see*.

The word *see* can be misleading. Trying to *see* a satellite is comparable to trying to see a dime from a distance of about a hundred miles. The word *see* in this context means there are no intervening objects between a satellite and a ground station designed to pick up satellite signals.

## CABLE AND SATELLITE TV

Cable TV services obtain their programs from satellites and deliver those programs to their subscribers via coaxial cable. The position of a cable company in the video hierarchy is that of a distributor. They must not only select television programs but must also process them so they can be utilized by their subscribers.

Since the cable company wires an entire community for video, it must amplify the signal so as to be able to satisfy the input signal level requirements of all the TV receivers connected to their system. In some instances, the cable company will encode (scramble) the signal so as to make it useless for any subscriber not paying for viewing a particular program. Such cable companies must supply decoders to their subscribers and this is usually done on a rental basis.

There are thousands of cable television companies scattered across the U.S. but these companies, supplying television programs to their subscribers via coaxial cable, prefer locations in areas of high population density. The cost of putting in cables, of labor, and of programs, means that substantial numbers of subscribers are needed to make such a venture profitable, particularly if the programs do not carry advertising.

While urban areas are successfully using broadcast and cable TV, there are tremendous sections of this country (and other countries as well) in which population density doesn't economically justify the installation of cable. Further, such areas may be outside the fringe limits of regular TV broadcasting, or, because of high terrain, may not have access to TV, even though the TV station or stations may be within 20 to 30 miles. For such locations, satellite TV is the only answer.

## THE LIMITATIONS OF CABLE TV

While cable companies do offer a greater program range than broadcast TV, they are limited in the variety and number of programs they offer. Users of satellite TV systems have no such limitations. With a clear view of the southern sky, they can select any satellite within the range of their satellite system and so have a substantially greater choice. Magazines are now published that supply the names and the times of transmission.

Since satellite programs are broadcast in real time, they can be time shifted by using a videocassette recorder, either VHS or Beta format, with a time shift capability. The setup is just the same as it is with the tape recording of terrestrial broadcast TV programs. For the owners of a satellite system, the amount of programming satellites can provide is enormous in comparison with VHF and UHF television broadcasting.

## COMMERCIAL MARITIME COMMUNICATIONS

Not all satellites are devoted to television broadcasting. Some are dedicated exclusively to commercial maritime communications and are known as Marecs and Intelsat V. The Intelsat V satellites are the largest of the commercial communications satellites. To get some concept of their size, the spread of the solar array they use is equivalent to the height of a five story building, while the antenna is more than 20 feet above its base. Marine communications satellites have been in orbit since 1976, providing commercial services over the Atlantic and Pacific, and, since 1978, to all ships in the Indian Ocean.

Other commercial satellites, such as those used for communications, include Palapa B, an Indonesian satellite. Palapa is an Indonesian expression meaning *fruit of effort*. Some satellites (Fig. 1-11) are used for gathering weather information.

## SATELLITE TILT

As indicated earlier, the orbit of a satellite lies approximately in the equatorial plane, an imaginary flat surface cutting through the equator and extending out into space.

With the passage of time, the orbit of a satellite (Fig. 1-12) will assume a tilt with respect to the equatorial plane. Known as the tilt angle or orbital inclination, the result is that the orbit of the satellite changes slightly during a 24-hour period.

Orbital inclination can be controlled and is done so from the earth. The instructions transmitted from the earth cause small hydrazine-fueled thruster rockets on the satellite to fire in a carefully planned sequence.

Adjustments to the orbit of the satellite are known as *station keeping* and make use of the satellite's telemetry, tracking, and command (TT & C)

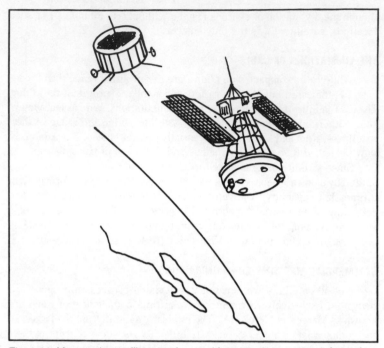

Fig. 1-11. Man-made satellites can be used for gathering weather information.

subsystem. Station-keeping thrusting maneuvers are also needed from time to time to maintain a satellite's east-west position relative to the earth.

On earth, an object that is made to move must overcome the forces of friction and gravity. These aren't serious factors in the airless, microgravity environment in which satellites travel. However, satellites are subject to the *push* of solar radiation. On earth, such a push would be of no consequence and would either not be noticed or would be ineffective. In the near zero gravity of space, the minute force of solar radiation can be a factor.

The satellite can be made to move along any one or more of three axes, each of which is controlled independently. Satellites also make use of gyroscopes that supply a counter torque to any deviation from a previously determined satellite position. These are supplemented, when required, by thrusters.

## SATELLITE POWER

All satellites contain receivers and transmitters (transponders) and, since these are active devices, require a source of DC (direct current) operating power. This is supplied, by batteries, but these require recharging to remain effective.

18

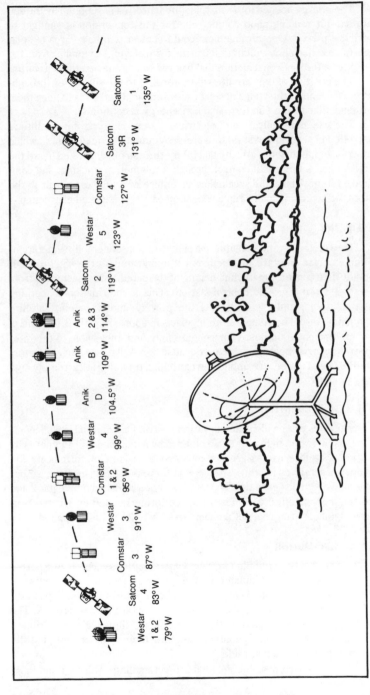

Fig. 1-12. Looking at the satellites from earth.

The energy needed for recharging the batteries is supplied by the sun with satellites using photovoltaic cells for the conversion of sunlight to electrical power. One arrangement could consist of a large area of solar cells (also called photovoltaic cells) in the form of wings extending from the satellite, a form of construction that has led to naming satellites as *birds*.

The position of the satellite with reference to the sun varies throughout the day and so the wings of solar panels are adjusted so as to keep them perpendicular to the sun for maximum energy reception.

The use and application of electrical power on board the satellite is controlled in several ways: (a) by a pre-programmed arrangement in which power is supplied automatically and (b) by the use of sensors on board the satellite and (c) by earth control through a telemetry, tracking, and command system. Thus, the weakening or failure of a unit can result in the automatic switching in of a replacement by earth transmitted instructions.

## TELEMETRY

The telemetry, the transmission of instructions for satellite operation, is duplex. Not only are instructions communicated to the satellite, but specific data concerning the functioning of the satellite is transmitted from the satellite to a ground control station. This is information concerning position, fuel supply, condition of the power supply system, satellite movement, etc. The temperature in space is approximately 3 Kelvins, a temperature some on-board components may find impossible to tolerate, thus requiring a heating system. Because air resistance isn't a factor, satellites need not be streamlined, even though traveling at extremely high rates of speed.

## SATELLITE RECOGNITION

Satellites can be identified in various ways. One method is alphanumerical, as indicated in Fig. 1-13. Another is by name, and still another is by location. Satellite names are accompanied by a number to indicate the quantity of satellites in orbit. The fact that there is a listing for Westar 2 but none for Westar 1 means that Westar 1 has been replaced by Westar 2 and that this new satellite occupies the space formerly used by Westar 1. In some instances, the channel number, from 1 to 24 is also supplied.

## SIGNAL SCRAMBLING

Signals can be scrambled in many ways. Picture synchronizing pulses can be weakened or eliminated so the resulting picture rolls, tears, or jitters. The polarity of the video signal can be inverted so the black portions of the picture become white and the white portions are viewed as black. The result is a television picture that resembles a film negative. Primary television colors can also be altered to show their complements, but the resulting picture is unviewable.

There are two possible solutions to scrambling. The first and most

| Designation | Satellite | Location |
| --- | --- | --- |
| F-5 | Satcom 5 | 143° W |
| G-1 | Galaxy 1 | 135° W |
| F-3 | Satcom 3R | 131° W |
| D-4 | Comstar D-4 | 127° W |
| W-5 | Westar 5 | 123° W |
| F-1/2 | Satcom 1/2 | 119° W |
| A-2/3 | Anik 2 & 3 | 114.5° W |
| AB | Anik B | 109° W |
| AD | Anik D | 104° W |
| W-4 | Westar 4 | 99° W |
| D-2 | Comstar 2 | 95° W |
| W-3 | Westar 3 | 91° W |
| D-3 | Comstar 3 | 87° W |
| F-4 | Satcom 4 | 83° W |
| W-2 | Westar 2 | 79° W |

Fig. 1-13. Satellite designations.

obvious is to become a subscriber, assuming, of course, the program producers will accept payment from the owner of a satellite system. They could consider satellite TV reception as competitive with cable, and yet only a small fraction of the country is wired for cable, nor does it seem that this small fraction will ever be increased substantially. As indicated earlier, cable TV is dependent on high population density.

An alternative is to build or buy a decoder, sometimes called a black box or unscrambler. These are being advertised and sold via newspapers and magazines. The legality of their use is a matter for the courts to decide.

In the meantime, very few satellite programs are scrambled, and the reason isn't hard to find. A single satellite can supply television programs to hundreds of cable companies, but not all of these want scrambling since it is expensive to supply thousands of subscribers with black boxes. Many potential subscribers would rather do without cable than to bear the expense of a decoder. Further, with decoders so readily available, cable subscribers can easily evade the monthly charge for a black box. Finally, encoding and decoding can degrade the quality of TV signals.

Even if satellite signals are scrambled, it does not follow that all the transmitters on that satellite make use of encoded signals. In some instances, the company responsible for program material will cooperate with TVRO (television receive only) owners. Thus, some of the channels on the satellite called Anik D1 are scrambled but the program operators, Canadian Satellite Communications, Inc., better known as Cancom, will either sell or lease unscramblers to TVRO users.

There is always the possibility that some of the more popular programs will be scrambled. However, it is equally possible that the number of scrambled programs will be kept to a minimum since encoding is expensive both for the programmer and the cable system operator. Most services do not consider private satellite receiving systems as a threat, because the majority of such installations are in areas not feasible for cable.

Cable companies not only supply programs that reach them via satellite, but are legally required to include programs generated by local television stations. For those who can pick up VHF and UHF television signals this is a duplication.

## DOMSAT

Not only the U.S. but a number of foreign nations also make use of satellites. To distinguish domestic satellites from foreign, they are known as domestic satellite transmissions. From this we promptly coin an acronym and call it Domsat.

## SATELLITE PICTURE QUALITY

With regular earthbound TV broadcasting, the signal must sometimes fight its way to get from the transmitting antenna to your home. Every hill or mountain is its enemy. So is every large building or bridge. In striking these structures, the TV signal is literally bounced—that is, reflected from these large surfaces. This reflected signal can still make its way to television receiving antennas, even though it is a bit delayed. Thus, it is possible for two signals from the same station to arrive at the same antenna, with one of the signals slightly delayed in time. As a result, two pictures are produced, one of which is a ghost, a picture somewhat weaker than that produced by the direct, unreflected signal, and displaced somewhat to the right. Depending on the number of reflections, there can be multiple ghosts, a series of pictures displaced to the right of the main picture. Satellite TV doesn't have this problem.

The earthbound broadcast TV signal has other enemies, devices that produce electrical noise. These behave like TV signals, but instead of enhancing the picture, they distort it in a variety of unpleasant ways. Satellite TV is far less susceptible to this kind of electrical noise.

Finally, satellite TV has the potential for supplying pictures having much higher definition. Better picture definition means more picture information must be transmitted. The frequencies used by satellite TV permit experimentation with high definition TV.

## UPLINK AND DOWNLINK SIGNALS

Satellites do not produce television pictures. Instead, the TV programs are produced on earth, are beamed up to a selected satellite (Fig. 1-14) and are then retransmitted to earth. The TV signal transmitted to the satellite is called an uplink signal; that from the satellite to earth is known as the downlink signal (Fig. 1-15).

## DUPLEX OPERATION

To keep the uplink signals from interfering with the downlink signals, duplex operation is used, which is a technical way of saying that the uplink and downlink signals work on different frequencies. The uplink frequency

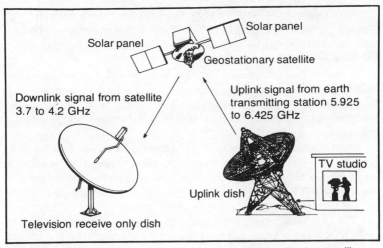

Fig. 1-14. Programs originate in TV studio and are transmitted to satellite.

band is between 5.9 and 6.4 gigahertz (abbreviated as GHz) but more precisely is 5.925 to 6.425 GHz. The downlink frequency range is between 3.7 GHz and 4.2 GHz.

The round trip distance between the earth and a satellite is 44,600 miles. Since the earthbound station producing the satellite TV signal could be just 50 miles or so from your home it does seem a roundabout way of delivering the signal to you. It is, but since television signals travel with the speed of light, or 186,000 miles a second, the time difference between direct TV reception and satellite TV reception is inconsequential.

## FIXED VS MOBILE UPLINK TRANSMISSIONS

In the early days of satellite TV, uplink signals were transmitted by fixed position earth transmitters. However, uplink signals can now be produced by mobile units, thus supplying on the spot broadcasts.

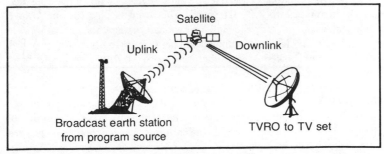

Fig. 1-15. Earth station sends program to satellite. Transponder in satellite routes signal back to earth.

Video signals can be transmitted to a satellite one at a time or more than one by using a special circuit arrangement known as a signal combiner.

## TRANSPONDERS

From the very beginning, starting with Sputnik I, satellites were considered more than just objects hurled into space. They were working devices and were expected to earn their keep, either in terms of scientific research, the transmission of data, worldwide telephone communications or entertainment.

A satellite, then, does more than just receive TV signals from earth. It converts the uplink signals it receives to downlink signals. This means each satellite has a receiver and a transmitter, a combined device known as a transponder. The satellite is also equipped with special antennas for receiving and transmitting the television signals.

The downlink signal, the signal transmitted by a transponder on a satellite, works under difficult circumstances. The transponder's electrical operating power is supplied by batteries that are kept charged by solar panels. Solar panels are transducers, converting one form of energy to another. Solar panels take the energy of the sun and convert it to electrical energy, but the conversion process is quite inefficient. Although, when extended, the solar panels of a satellite are the largest part of that satellite, they manage to produce just enough energy so that approximately 5 watts is the maximum transmission power of each transponder.

This is the average transmission power of many satellites, but some have more by using various expedients. Some have more efficient solar panels, while others use fewer transponders. A typical satellite might have 24 transponders. Instead of distributing the available electrical energy among so many, some satellites have only 12. Some, although they have 24 transponders, may use these on a staggered basis. They may have an arrangement in which not all of the transponders work at the same time.

Transponders, of course, could use a greater amount of power, certainly much more than 5 watts. The technology for supplying a much higher operating power and the equipment for supplying that power is available. However, the frequency band used by the satellites for downlink transmission is in the 3.7 to 4.2 GHz band. This band is also the microwave frequency range used by the Bell Telephone Company for the nationwide transmission of telephone communications. Consequently, satellite power in the band must be kept low, deliberately so, to avoid interference with our transcontinental telephone service.

The amount of available operating power to be used by a transponder is limited by two factors: the first is the amount of electrical charging power made available by solar cells, and the other is the ampere-hour capacity of the batteries. The solar cells keep the batteries charged, but solar cells are comparatively inefficient.

If fewer transponders were used, a number less than 12 or 24, the operating power to the remaining few transponders, possibly a half dozen or less, could be increased considerably. There is no choice available to the operators of the satellite transponders since they have no option but to avoid interfering with terrestrial microwave telephone communications.

While a typical transponder has an input power of about five watts dc to its final power amplifier stage (the stage that delivers signal power to the antenna), there is one satellite that exceeds this. This the the Anik D satellite whose orbital position is 104.5 degrees west. The 24 transponders on this satellite have an operating power of 11.5 watts. There are 24 transponders using dual polarity, that is, the transmitted signals are vertically and horizontally polarized. Anik satellites are Canadian.

If programs other than video are to be transmitted, such as voice communications which require considerably less bandwidth, the total number of transponders could be more than 24. Transponders handle a variety of signal sources other than video and telephony. This includes broadcasts of sound in both mono and stereo, and the transmission of data and news.

## TELEVISION CHANNELS

In a sense, a satellite could be compared to a television station, since it transmits television signals. Unlike a television station which broadcasts just a single channel, a satellite can use each of its transponders for that purpose. Thus, if a satellite has 24 transponders it has the capability of transmitting 24 programs, and can do so either simultaneously or sequentially. Each of these programs is a channel, and so just a single satellite can transmit twice as many channels as the combined low frequency and high frequency VHF television bands. Identically numbered channels all use the same operating frequency, even if supplied by different satellites.

## MOBILE UPLINK

Earthbound TV broadcast stations use portable equipment for electronic news gathering (ENG). Similar techniques are used for satellite transmissions. The amount of power required by an earth station for sending a TV signal to a satellite is quite small, particularly when compared to the amount of power used by TV broadcast stations. Consequently, on-the-spot coverage is available for satellite use. News reporters are equipped with portable units that transmit the signal to a nearby mobile station which then transmits the video signal and sound to its selected satellite.

## THE C BAND

Although there are uplink and downlink signals, our only concern in operating a satellite receiving system is those that are downlink. These, as indicated earlier, are in the frequency range of 3.72 GHz to 4.18 GHz, a band that is sometimes designated by the letter C.

The C band is not the only one set aside for the reception of satellite

TV signals. There is still another band of even higher microwave frequencies, ranging from 11.7 to 12.2 GHz, known as the Ku band, or sometimes simply referred to as the K band. This band of frequencies is reserved for direct broadcasting satellite services, commonly known as DBS. DBS is described in more detail in a separate chapter, Chapter 8.

## THE COMMON CARRIER CONCEPT

The owners of satellites are large corporations, and they are legally permitted to rent the services of their transponders. In so doing, they act as common carriers with the companies renting the transponders responsible for supplying program material.

Not all of the transponders are offered on a common carrier basis. In at least one instance, there is private ownership of a transponder, and so a company that owns a satellite may elect to sell its transponders, instead of renting them.

## UPLINK POWER

Compared to TV broadcast stations, transmitters used for sending uplink signals operate with modest amounts of power. This power is in the range of 1 to 3 kilowatts. While this isn't all that much (a toaster can easily use 1 or more kilowatts and so can an electric heater) the transmitting dish used is quite large, often having a diameter of as much as 30 feet. Since this power is radiated directly into space and since it is reflected and concentrated prior to transmission, the signal energy reaching the satellite is practically free of noise, that is, it is a so-called *clean* signal.

The satellite does not improve the signal. All it can do is receive the TV signals beamed up to it from earth, and downconvert them, that is, change them from the higher uplink frequency to the lower downlink frequency.

The downconversion process does not *clean up* or modify the signal in any way. The only purpose of downconversion is to permit duplex operation.

## TRANSPONDER TEST PATTERNS

In the early days of TV, terrestrial stations would broadcast a test pattern for some time prior to televising actual program material. The structure of the test pattern was such that it enabled television technicians to make adjustments to receivers for best reception. Television receiver circuitry, however, has now advanced to the point where such adjustments are no longer required. Consequently, the use of test patterns for ground broadcast TV has practically disappeared.

Transponders now use test patterns also. Known as test cards, these are quite unlike the test patterns used three or four decades ago by earth TV broadcast stations. Test cards sometimes supply operating information, while others simply seem decorative. They do have one advantage. Since they supply a still picture, it is convenient to use them for making adjustments to a satellite receiving system.

## EFFECTIVE ISOTROPIC RADIATED POWER

Electrical energy is simply electrical power multiplied by the length of time it is used. Electrical power is delivered to your home via lines from your local power utility. Electrical power is also delivered to your home from a satellite through space. In either instance electrical energy is involved, although the electrical energy from the satellite is extremely small in comparison.

Effective isotropic radiated power (EIRP) is a measure of the relative strength of the satellite TV signal for different geographical areas and is supplied in terms of decibels (dB) with reference to 1 watt. A characteristic of EIRP is that the signal strength of a satellite can be plotted in the form of contour lines, as shown in Fig. 1-16.

## FOOTPRINT

The diagram in Fig. 1-16, known as a footprint, indicates that the greatest signal strength is in the area around the center of the inner contour line. The diagram indicates numbers: that at the center is 34, and that outlining the continental U.S. is 32. These numbers indicate signal strength, in decibels, referenced to one watt, and is written as dBw. Thus, the center-contour is 34 dB, with 1 watt used as the reference point. This does not mean the center contour is 34 watts.

The footprint pattern is not the same for all satellites and often the pattern will be different for the various transponders on the same satellite. The footprint shown in Fig. 1-16 is for Satcom's satellite F3 positioned at 131 degrees west longitude. For this particular satellite, the EIRPs for transponders 3, 7, 11, 15, 19 and 23 will be at least 2 dB higher than those shown.

The boresight point for the footprint in Fig. 1-16 is Denver. In that city the EIRP is 34.8. The EIRP for Anchorage, Alaska is 28.0 to 28.7; for Juneau, Alaska it is 30.4 to 31.3 and for Honolulu, Hawaii it is 25.5.

## BORESIGHT POINT

All receiving and transmitting antennas have a radiation pattern, indicating the areas over which these antennas are most effective. The pattern could be a series of concentric circles, or ellipses, or some geometric pattern between these two.

The center of such a pattern is the point of maximum signal strength and is referred to as the boresight point. The boresight points of each satellite signal do not necessarily correspond. The illustration in Fig. 1-17 is the footprint for RCA's Satcom 1 satellite with the numbers indicating relative values of signal strength. The contour lines indicate that this

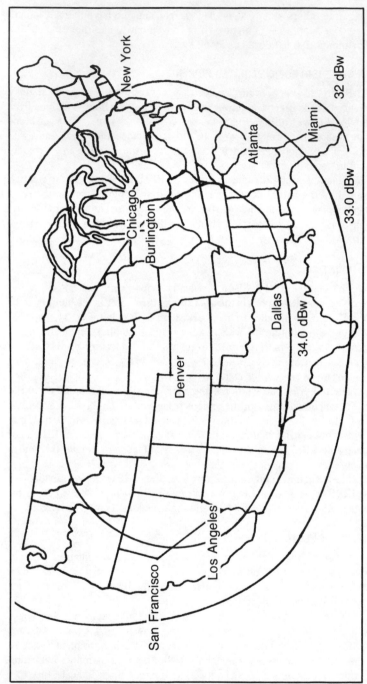

Fig. 1-16. Footprint of Satcom F-3R.

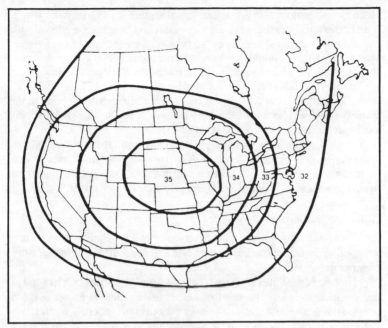

Fig. 1-17. Footprint for Satcom 1. (RCA)

particular transponder's signals reach the earth somewhat comparable to a beam of light from a flashlight.

The footprints indicate that not all signals received over a wide geographic area will have the same strength. For the continental U.S., the signals will range from 30 to 37 dBw. Generally, you will receive all the channels from a particular satellite with equal quality but it is possible for one or more channels to be a little less satisfactory than the others.

With approximately 5 watts of output power for each transponder, the signal arrives on earth having lost approximately 196 dB of its signal strength. This means that the signal strength has dropped from 5 watts at the transponder to $0.5 \times 10^{-20}$ watt. That number, $-20$, is known as an exponent and is a convenient way of writing the number 1 preceded by 19 zeros. To really get the inpact of the drop in signal strength of the TV program as it races through space from the satellite to the home receiver, we are comparing 5 watts to 0.000000000000000000005 watt. In short, the signals that reach the earth from the satellite are practically non-existent. In contrast, VHF broadcast television signals that reach TV antennas are often a million (or more) times stronger than satellite signals.

## SATELLITE ANTENNAS

A satellite can beam its signals toward earth using either wide beam or narrow beam antennas. The wide beam antennas, known as global beam or broad beam, could direct a satellite signal so that it covered more than a third of the surface of the earth. By using just three satellites equipped with global beam antennas, the entire surface of the earth could receive the signals. With the satellites positioned above the equator, as they now are, about the only regions on earth not covered are the North and South Poles.

Satellites, however, use narrow beam antennas. These are more directional and are characterized by high gain. High gain does not mean the antenna amplifies, for the antennas are passive devices. High gain is simply a comparison with a reference standard.

Narrow beam antennas are practically essential for transponders, since their power output is so small. A narrow beam antenna concentrates its transmitted energy; a wide beam antenna does not. Even with narrow beam transmission, a single transponder easily covers the U.S.

## CROSS POLARIZATION

To avoid interference with each other, television stations on earth are assigned different operating frequencies. This is also true of satellite transponders.

However, satellites commonly have as many as 24 transponders and to minimize interference from one to the other use a technique known as cross or opposite sense polarization. An easy way to consider cross polarization is to think of each satellite signal as traveling to earth in a horizontal flat plane. All signals traveling along similar flat planes are said to be similarly polarized. If a transponder now transmits a signal that is at right angles to the original flat planes, this next set of signals could be called vertically polarized. Because the signals are vertically and horizontally polarized, they tend not to interfere with each other.

Using horizontal and vertical polarization of signals means that each satellite can double the number of transponders it can use. Instead of 12, it can have 24.

The polarization process isn't a haphazard one. Thus, all odd numbered channels could be vertically polarized; all even numbered channels horizontally polarized (Fig. 1-18). With this arrangement no two vertically or no two horizontally polarized signals are transmitted in adjacent channels.

A receiver tuned to one channel could possibly respond to other channels on either side of the selected channel. The polarization technique

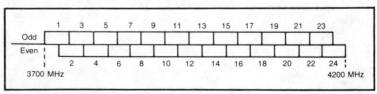

Fig. 1-18. Channel arrangements for transponders having horizontal and vertical polarization.

| Channel number | Frequency (MHz) | Satcom 1, 2, 3R & 4 (Polarization) | Comstar 1, 2, 3, & 4 (Polarization) | Westar 4 & 5 Anik D1 (Canada) (Polarization) | Westar 1, 2, & 3 Anik B (Canada) (Polarization) |
|---|---|---|---|---|---|
| 1 | 3720 | 1(V) | 1V(V) | 1D(H) | 1(H) |
| 2 | 3740 | 2(H) | 1H(H) | 1X(V) | |
| 3 | 3760 | 3(V) | 2V(V) | 2D(H) | 2(H) |
| 4 | 3780 | 4(H) | 2H(H) | 2X(V) | |
| 5 | 3800 | 5(V) | 3V(V) | 3D(H) | 3(H) |
| 6 | 3820 | 6(H) | 3H(H) | 3X(V) | |
| 7 | 3840 | 7(V) | 4V(V) | 4D(H) | 4(H) |
| 8 | 3860 | 8(H) | 4H(H) | 4X(V) | |
| 9 | 3880 | 9(V) | 5V(V) | 5D(H) | 5(H) |
| 10 | 3900 | 10(H) | 5H(H) | 5X(V) | |
| 11 | 3920 | 11(V) | 6V(V) | 6D(H) | 6(H) |
| 12 | 3940 | 12(H) | 6H(H) | 6X(V) | |
| 13 | 3960 | 13(V) | 7V(V) | 7D(H) | 7(H) |
| 14 | 3980 | 14(H) | 7H(H) | 7X(V) | |
| 15 | 4000 | 15(V) | 8V(V) | 8D(H) | 8(H) |
| 16 | 4020 | 16(H) | 8H(H) | 8X(V) | |
| 17 | 4040 | 17(V) | 9V(V) | 9D(H) | 9(H) |
| 18 | 4060 | 18(H) | 9H(H) | 9X(V) | |
| 19 | 4080 | 19(V) | 10V(V) | 10D(H) | 10(H) |
| 20 | 4100 | 20(H) | 10H(H) | 10X(V) | |
| 21 | 4120 | 21(V) | 11V(V) | 11D(H) | 11(H) |
| 22 | 4140 | 22(H) | 11H(H) | 11X(V) | |
| 23 | 4160 | 23(V) | 12V(V) | 12D(H) | 12(H) |
| 24 | 4180 | 24(H) | 12H(H) | 12X(V) | |

H-Horizontal V-Vertical

Fig. 1-19. Channels, downlink frequencies, and polarization of 14 satellites.

results in an attenuation of some 30 dB, a ratio of 1,000 to 1. This means the channel chosen for reception will be a thousand times stronger than any adjacent unwanted channel. The advantage of the polarization technique is fairly clear. It makes much more effective use of available channel space, in effect letting us have two channels for each one not polarized.

Not all satellites take advantage of horizontal and vertical polarization of downlink transmissions. Thus, Westar 1, Westar 2, Westar 3 and Canadian satellite Anik B use horizontal polarization only. (Anik is an Eskimo word meaning *little brother*.) Most satellites, though, use both forms of polarization, including Satcom 1, 2, 3R and 5; Comstar 1, 2, 3 and

4; Westar 4 and 5 and Anik D1. The chart in Fig. 1-19 shows the downlink frequencies of these satellites. Those using one type of polarization have only 12 downlink frequencies supplied by 12 transponders. Those using both vertical and horizontal polarization use 24 transponders; 12 vertically polarized; 12 horizontally polarized.

## TRANSPONDER OPERATING FREQUENCIES

Figure 1-20 is a list of downlink frequencies used by a satellite having 24 transponders. These are exactly the same frequencies used by all the other video satellite transponders. In other words, all the satellites operate on the same frequencies.

It would seem, offhand, that since the transponders of the different satellites use the same frequencies, there would be considerable difficulty with interference. However, there are several factors that minimize or prevent this. Since it is possible to *tune the satellites*, that is, select just a single satellite for reception, TV channels supplied by all other satellites are rejected. The concept is the same as tuning in one particular channel on VHF TV. Further, cross channel pickup is reduced through the use of alternate vertical and horizontal polarization, as described earlier. The design of the dish, the reflector used for concentrating the signal strength of a selected satellite, is an important factor in minimizing possible signal interference from another satellite during downlink transmission (Fig. 1-21).

## TRANSLATION FREQUENCY

The uplink signals to a satellite can have a frequency in the range from 5,925 MHz (5,925 megahertz or 5.925 GHz) to 6,425 MHz (6.425 GHz).

| Ch | Freq | Ch | Freq |
|----|------|----|------|
| 1 | 3720 | 13 | 3960 |
| 2 | 3740 | 14 | 3980 |
| 3 | 3760 | 15 | 4000 |
| 4 | 3780 | 16 | 4020 |
| 5 | 3800 | 17 | 4040 |
| 6 | 3820 | 18 | 4060 |
| 7 | 3840 | 19 | 4080 |
| 8 | 3860 | 20 | 4100 |
| 9 | 3880 | 21 | 4120 |
| 10 | 3900 | 22 | 4140 |
| 11 | 3920 | 23 | 4160 |
| 12 | 3940 | 24 | 4180 |

Fig. 1-20. Transponder channels and their downlink frequencies.

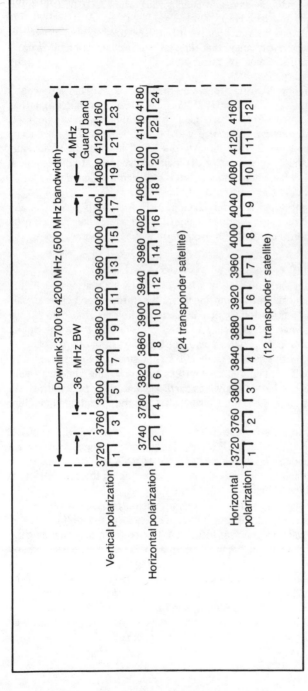

Fig. 1-21. The bandwidth of a satellite transmitted signal is 36 MHz. This bandwidth includes both the video signal and audio subcarriers. The 36 MHz wide signal is centered in the transponder, permitting a 2 MHz guard band at the upper and lower frequency limits. In a 24 channel system, odd numbered channels use the same channel center frequencies and bandwidth as the transponders 1 through 12 in a 12-channel system. The even numbered Channels 12 through 24, are located between the odd numbered Channels with a 20 MHz center-to-center spacing between adjacent channels. Cross polarization, also called dual polarization, places the downlink signals of alternate channels at right angles to each other. (M/A-Com Video Satellite, Inc.)

33

The downlink signals can have a frequency range from 3,720 MHz to 4,180 MHz (3.72 GHz to 4.18 GHz.).

If a satellite has 24 transponders it will have a 24 channel capability. This doesn't necessarily mean that television programs will be broadcast on all 24 channels. Some of the channels may not be used for video entertainment purposes.

The frequency difference between an uplink signal and a downlink signal is constant and is 2,250 MHz (2.25 GHz). Thus, if an uplink signal has a frequency of 5,970 MHz its corresponding downlink signal will be 3,720 MHz and the frequency difference will be 2,250 MHz. This frequency difference is known as the translation frequency. The translation frequency is the result of the downconversion of the uplink frequency to the downlink frequency. In effect, every uplink signal is reduced in frequency by 2,250 MHz prior to being transmitted to earth.

The example supplied is for Channel 1 (Fig. 1-20). If you select any other channel at random, possibly Channel 12, with a downlink frequency of 3,940 MHz, then the translation frequency will be the same.

## AN OVERALL VIEW

Basically, a satellite is just a *housing* for relay stations receiving signals from earth and retransmitting them. Satellites receive their signals from satellite uplink stations which, in turn, get their program material via cable or microwave, or by using a combination of the two. The uplink station may receive its program material from a variety of sources.

To check on the quality of their transmissions, uplink stations are equipped not only with uplink transmitters but a downlink system as well. Uplink stations are better known as *earth stations* and, by definition, an earth station is engaged in the transmission and the reception of satellite signals.

The transponders on satellites, then, are simply television signal sources, just as a VHF or UHF terrestrial station is a signal source, or a video disc or a video tape. In the case of terrestrial TV, all that is required is a television set and an antenna capable of picking up earthbound television stations. Satellite TV is somewhat different.

A system used only for receiving television programs from one or more satellites is called a television receive only system, abbreviated as TVRO. An overall view, such as in Fig. 1-22 is helpful in understanding the relationship of the various components in a TVRO system to each other.

The system starts with a dish for reflecting the downlink signals to a feed; a waveguide that delivers the television signal to a small probe; the antenna. From the antenna, the signal is routed, via coaxial cable, to a low-noise amplifier (LNA), generally mounted on the dish. The output of the LNA is supplied to a downconverter, a device that works as a frequency changer. The downconverter is the first step taken to reduce the frequency of the carrier from 4 GHz.

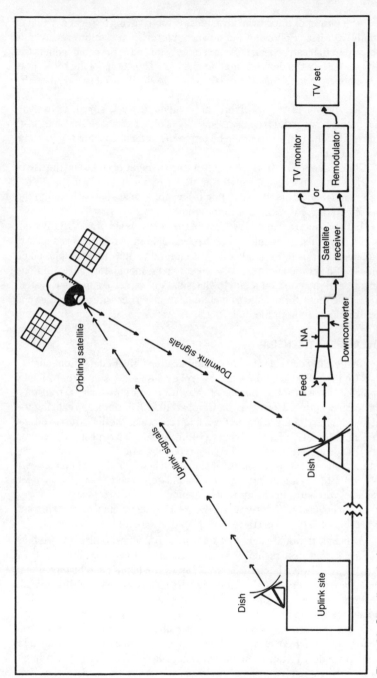

Fig. 1-22. Signal's path, from sender to receiver. (McCullough Satellite Equipment, Inc.)

The output of the downconverter is a wave having a frequency that is usually 70 MHz. Following the downconverter, is a satellite receiver, a component that can be manually or remotely turned. This component is the unit that selects the desired satellite channel. The output of the receiver is the composite video signal, similar to that originally used to modulate the original uplink wave.

The output of the receiver, the composite video signal, is used to modulate a new carrier having a frequency of either Channel 3 or Channel 4. This modulated carrier can then be fed to the antenna input of the in-home TV receiver.

It may seem that this represents a large number of steps and the use of a variety of components, but the thing to remember is that the TVRO system must adapt itself to existing television signal standards and to the use of existing TV receivers in the home.

In the chapters that follow, each of the components of a TVRO system will be examined in detail. These are available in two forms: as separate components or as integrated units. An integrated unit, for example, could consist of a downconverter, a receiver, and a modulator, all within one console and mounted on a single chassis. A separate component system might have these three units as individual items. There are advantages and disadvantages to both methods.

## GOVERNMENT LICENSING

In the early days of satellite communications, the Federal Communications Commission maintained full control over satellite signal reception by requiring licenses for all phases of satellite operation—that is, transmission and reception. However, on October 10, 1979, perhaps realizing the dubious legality of insisting on licensing reception, the FCC removed the licensing restriction for receiving satellite signals. This means anyone can install and operate a satellite signal receiving station for personal, not commercial, use. The reception of signals, whether radio as in the case of AM and FM broadcasting, or earthbound television, has always been available without a licensing requirement.

Consequently, according to the FCC, the manufacture, sale, and ownership of satellite earth receiving stations are all legal activities.

Offhand, it would seem that terrestrial TV and satellite TV are both broadcast, and in a technical sense they are. However, terrestrial TV signals have always been made available to the public, and further, these signals are not aimed or directed at a particular antenna. Satellite signals are broadcast only in the sense that they are transmitted, but satellite signals are intended for pickup by specific dishes, hence they are sometimes referred to as non-broadcast signals. Oddly enough, terrestrial broadcasts do not cover as much geographical area as satellite transmissions. A single TV station might supply signals suitable for a part of a single state. A single satellite signal could reach most of the U.S.

According to the FCC: "We are not aware of any section 605 (Communications Act of 1934) criminal or civil cases involving earth stations."

The FCC is an arm of the Federal government. The installation of a dish may involve local ordinances and these can vary from one municipality to another. A frequent concern of local government is that the dish should not be aesthetically disturbing and should not result in a hazardous condition. Some municipalities may require a license for installation; most do not.

## THE PROBLEM OF COPYRIGHT

At the present time the question of copyright is very much *up in the air* and is being debated in the courts. If your satellite system picks up programs and you watch these programs in the privacy of your home you do not need to pay a fee or obtain a license of any kind. However, if you plan to go into the business of taping off the air, and selling duplicates of the tapes you may well be in legal difficulties. The 1979 ruling of the FCC is interpreted to mean that not only individuals can use TVROs for in home television viewing, but commercial enterprises such as hotels, motels, bars, hospitals, high-rise apartment houses, and condominiums.

## USES FOR SATELLITE TV

While there is considerable emphasis on the use of satellite TV as a means of video entertainment, it is less generally known that satellite TV has considerable potential in other areas. It can be used to supply an in-home hi/fi system with stereo sound, something that terrestrial broadcast TV is currently not doing. It has substantial possibilities in education and it is proving its worth in teleconferencing. Satellite stereo audio is described in some detail in Chapter 6.

### Applications in the Schoolroom

There are two primary classroom uses for satellite communications. The first is classroom viewing. This can mean conventional viewing of live programming with its immediacy, currency, and topicality. However, live material can also be recorded for subsequent individual or group instruction.

Satellite programs can be stored on tape for immediate or time-shift viewing. Educators can choose program material from conventional private and educational channels or from dedicated channels such as Learn Alaska, the Learning Channel or Western ITV, which are devoted exclusively to educational purposes. According to Dr. Robert H. Decker, Director of Satellite Education for KLM Electronics, Inc. "Satellite technology can enable classroom teachers to encourage pursuit of a course of study. It will let educators incorporate in their lesson plans an in-depth visual perspective on a variety of concepts.

An example is a biology lab where instructional satellite programs can dramatize the procedure for making incisions. Satellite TV can present accurate perspectives on childbirth, major surgical techniques, the analysis of stress points in wind-tunnel testing, cultural nuances among aborigines, the behavior of wildlife . . . the possibilities are endless."

## Satellite Teleconferencing

Video teleconferencing permits simultaneous transmission of audio/visual material to multiple locations where students view proceedings on monitors or large screens and interact with teachers and other students via phone links.

The National University Teleconferencing Network operates teleconferences for 66 colleges and universities in 38 states plus the Smithsonian Institute in Washington, DC. The Public Service Satellite Consortium (PSSC) is an international non-profit organization with more than 100 member groups in education, health, medicine, public broadcasting, libraries, state telecommunications, and religious communications.

Duke University Medical Center's television earth station permits doctors to be involved in nationwide medical research projects and study new patient care technique without leaving the hospital, at great savings in travel time and dollars.

On September 7, 1983 the National Diffusion Network, an agency of the U.S. Department of Education, sponsored a national educational teleconference for teachers, the first of its kind. The NDN teleconference was beamed to all 50 states.

Studies by Bell Telephone and the PSSC suggest that 35% of travel to meetings can be eliminated by teleconferencing. *Satellite Orbit* magazine reported that the University of Mississippi had its first satellite dish installed in 1982. Dr. Will Norton, chairman of the department of journalism has stated, "I don't doubt that every university building in the country will one day have a satellite dish."

# Chapter 2

# Signal Processing

Every television program starts with two transducers: a microphone for the conversion of sound into an equivalent electrical voltage and a video camera for the conversion of pictures, also into an equivalent electrical voltage.

With these two voltages on hand the next step is to transmit them to a selected satellite, and from the satellite to TVROs. The final step, of course, is delivery to the antenna input terminals of a television set. To achieve this up and down transportation to and from a satellite, and from the dish to the television set, the two voltages, audio and video, must be processed.

## DEVELOPMENT OF A CARRIER WAVE

By themselves, audio and video signals are ill equipped to travel, for neither are capable of moving any great distance. To enable them to do so, we must load them onto another wave, aptly named a carrier. The carrier is simply a means of transportation, but not all carriers are alike for they can and do differ in frequency, and consequently in wavelength as well.

## THE MEANING OF FREQUENCY

Frequency is a commonly used word and all it means is the number of complete waves per unit of time; a second. Figure 2-1 is a representation of a single complete wave. If it takes one second from the start of the wave to its finish, the frequency is one cycle per second (one cps), more commonly known as a hertz, abbreviated as Hz.

The voltage delivered to the power outlets in your home has an appearance similar to that in Fig. 2-1 except that there are many more

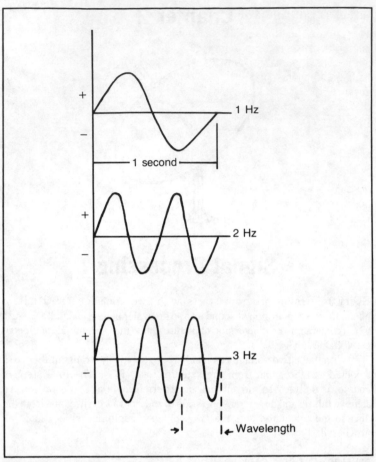

Fig. 2-1. The higher the frequency, the greater the number of wave cycles per second. A gigahertz has 1,000,000,000 of these per second. The waves are produced by electronic oscillator circuits. Wavelength is measured from beginning of one wave to the start of the next. As frequency increases, wavelength decreases.

complete waves per second, either 50 or 60. Thus, the line frequency can be said to be 50 Hz or 60 Hz.

## CONTINUOUS WAVES

One of the characteristics of the wave shown in Fig. 2-1 is that its amplitude is constant. Any one wave is identical in height to any other wave. Such waves are referred to as continuous waves or CW. Continuous waves are important because they are our signal carriers. The continuous waves delivered to your home electrical outlets are produced by mechanical generators. Those used as carriers are produced by electronic circuits.

## FREQUENCY NOMENCLATURE

To avoid the messiness of using large numbers, continuous waves have various names, depending on frequency. Thus, one thousand cycles per second, better known now as 1,000 Hz, is abbreviated as 1 kHz. A wave having a frequency of 1 kHz is one that has 1,000 complete waves per second. Since the time element, the second, must always accompany any discussion of frequency, it is generally "understood" and so is often omitted. Thus, a frequency may be indicated as 500 cycles but is 500 cycles per second or 500 Hz. A frequency of 10 kHz is 10,000 cycles, actually 10,000 cycles per second.

Frequency is also involved with larger numbers. Thus a million cycles per second is 1,000,000 cps, or using the preferred form, 1,000,000 Hz or 1 megahertz, abbreviated as 1 MHz. 20 MHz is equivalent to 20,000,000 cps.

The MHz abbreviation is commonly used in connection with the broadcast of TV and FM signals. Thus, the range of FM broadcast signals is from 88 MHz to 108 MHz.

In connection with the transmission and reception of satellite signals, another frequency abbreviation is used and that is the gigahertz, commonly written as GHz. A gigahertz is 1,000 megahertz or one thousand million cycles per second. This is a fantastic frequency for it means an electronic generator is producing 1,000,000,000 complete cycles per second, every second.

## WAVELENGTH

Wavelength is the distance from the start of a single wave to its end, as indicated in Fig. 2-1. Since, as frequency increases, we get more cycles per second, the wavelength decreases. Consequently, for frequencies in the megahertz region, wavelengths are extremely small, and become even smaller for frequencies in the gigahertz range. Frequency and wavelength have an inverse relationship. As frequency increases, wavelength decreases, and vice versa. Since the wavelengths of carriers used in satellite signal transmission and reception are so very small, they are sometimes referred to as microwaves. A microwave could be any wave having a frequency of 1 GHz or more. While there are no agreed upon rules governing the nomenclature of radio waves, the term *microwaves* is often applied to both uplink and downlink carrier waves.

One problem with using the word *microwave* is that it may possibly be misinterpreted. It does not mean the waves are small in size, for wavelength has nothing to do with the strength of a wave. All it does mean is that the waves are extremely close together, that the distance from the start of one wave to the start of the following wave is extremely short.

## QUASI-OPTICAL WAVES

As the frequency of a wave is increased, it begins to approach the frequencies of light, for these are also waves. As they do, they begin to take on some of the properties of light and are said to be quasi-optical. The microwaves used in satellite transmission can be focused, just like a beam of light. These waves also travel in straight lines. Thus, when transmitted in a horizontal plane from an earth station they would not follow the curvature of the earth, as the much lower broadcast radio frequencies do, but would travel in a straight line.

Like light waves, microwaves can be reflected and refracted. They can be focused so that they meet at a point or so they can be reflected in beam-like form. Because their wavelengths are so short, the antenna for responding to such waves is equally short, often no more than an inch in length.

## AUDIO AND VIDEO WAVES

While sound can be transduced into an audio wave and light reflections into a video wave, these two waveforms, audio and video are unlike continuous waves, although there are some points of similarity. First the differences, one of the most important of which is the fact that audio and video waves do not have a constant amplitude. They keep changing in strength from one moment to the next. Another is the fact that they do not constitute a single frequency but a group of frequencies. Thus, an audio wave is generally considered to be any frequency in the range of 20 Hz to 20 kHz. A video wave has an even wider range, from about 50 Hz to 4.2 MHz. Considering similarity to continuous waves, both audio and video waves can be specified in terms of frequency or wavelength.

## THE BANDWIDTH CONCEPT

Since audio and video waves are groups of frequencies instead of a single frequency, such as a continuous wave, the range of such waves is considered to be from the lowest frequency to the highest. This is referred to as its bandwidth. Thus, the bandwidth of audio is 20 Hz to 20 kHz and the bandwidth of video is 50 Hz to 4.2 MHz. The bandwidth of a video signal is much wider than that of audio since it represents much more signal information.

Just as an automobile occupies space on a road, so to do bands of frequencies require space. A band of frequencies is often referred to as a channel, with two frequencies supplied to indicate the upper and lower frequency limits of that channel.

## BANDWIDTH VS CHANNELS

Although these two words are sometimes used interchangeably, there is a difference. Thus, in the case of a television signal, the video portion occupies a space of approximately 4 MHz and the accompanying audio signal requires a space of 0.5 MHz, making a total of 4.5 MHz. The channel space allotted to this signal is 6 MHz. This is comparable to a vehicle that is 6 feet wide but requires a minimum road space of 8 feet.

The frequency range of the AM broadcast band extends from 535 kHz to 1605 kHz, or a total bandwidth of 1070 kHz, a little more than 1 MHz or 1,000,000 Hz. A single television signal for picture and sound requires a channel that is 6,000,000 Hz or 6 MHz. Thus, just one television channel requires a bandwidth that is six times greater than that of the entire AM broadcast band. There are 12 terrestrial television broadcasting channels.

Each AM broadcasting station requires enough space for 10 kHz, called a channel, to indicate our recognition that a band of frequencies (not just a single frequency) is involved. An AM station might be assigned an operating frequency of 550 kHz, but this frequency can extend from 545 kHz to 555 kHz during transmission, a total band spread or bandwidth of 10 kHz. Bandwidth indicates the maximum amount of air space a signal will occupy.

Bandwidth, then, is one of the important factors and differences between AM broadcasting and television broadcasting. Television broadcasting requires a tremendous amount of air space for each channel.

## THE SATELLITE BAND

The operating frequencies used by satellites are higher than those used by terrestrial TV and are measured in gigahertz. Television programs produced by earth stations for uplink use have frequencies between 5.9 and 6.4 gigahertz. This doesn't sound like much until you do a little arithmetic. A gigahertz is 1,000 MHz or 1,000,000,000 Hz. In going from 5.9 GHz we move from 5,900,000,000 Hz to 6,400,000,000 Hz, or, subtracting the smaller number from the larger, 0.5 GHz or 500 MHz. This means that the space assigned to satellite television encompasses a range of 500 MHz or 500,000,000 Hz.

Compared to the bandwidth allowed to AM broadcasting it is 500,000,000 compared to 1,070,000, not quite 500 times as much. The road leading from earth to the satellites is 500 MHz wide. If we were to use earthbound television channels, with each having a 6 MHz spread, we could accommodate 500 ÷ 6 or approximately 80 channels for each satellite. However, as you will see, this isn't done so as to be able to supply the satellites with signals that are as noise free as possible. In essence, that is one of the great advantages of working with gigahertz frequencies—there is more room for a large number of channels.

## THE NEED FOR BANDWIDTH

A signal that is to be transmitted can be regarded as information. A voice signal, music, or video, are various forms of information. Some information is much more detailed than others. Morse code, when air transmitted, is simply a sequence of dots and dashes. A binary signal is equally simple, consisting of a series of equally spaced pulses or absence of pulses. The waveforms representing Morse code or binary pulses aren't all that complex and so the bandwidth they require is limited. As we move toward voice transmission and video, the information becomes more detailed, and the need for bandwidth increases.

The human voice has a frequency range, generally from about 100 Hz to 3,000 Hz. The wave that is transmitted is a composite of many frequencies, all within this range. So the signal that is to be transmitted must have a bandwidth of at least this much.

Consequently, a *message* that has a substantial amount of information requires a wider frequency range than messages that have much less information. In terms of numbers, the channel width requirements for a telephone conversation is 3,000 Hz, for an FM channel 150 kHz and for a video signal, 6,000 kHz.

## THE NEED FOR MODULATION

We now have three types of waves that are involved in the delivery of a signal to a satellite and from a satellite back to a TVRO. These are a continuous wave, a video wave, and an audio wave. The continuous wave, known as a carrier, is used as the delivery method, since it is capable of traveling great distances. The process of loading the audio and video signals on the carrier is known as modulation. The audio and video signals supply information; the carrier supplies transportation. The two most common methods of modulation (there are others) are amplitude modulation (AM) and frequency modulation (FM).

## AMPLITUDE MODULATION

There are various circuits used for amplitude modulating a carrier wave, but the results are the same. Figure 2-2 shows a carrier wave, prior to modulation while Fig. 2-3 illustrates the same carrier wave after the audio signal has been loaded on it.

There are several things to note from these drawings. The first is that the frequency of the carrier wave has remained unchanged. What has

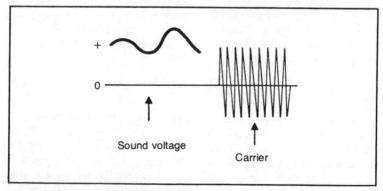

Fig. 2-2. Example of amplitude modulation. The sound voltage at the left is impressed on the much higher frequency carrier wave shown at the right.

44

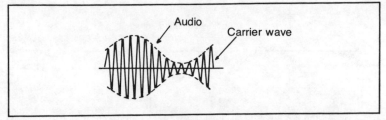

Fig. 2-3. In amplitude modulation (AM), carrier frequency remains constant but amplitude changes in step with the audio signal.

happened is that the amplitude of the wave has been modified and is now varying at an audio rate. Like the continuous wave, the modulated wave is capable of traveling greater distances, but in so doing, carries the audio wave along with it, *piggyback* so to speak. The circuit that loaded the audio wave on the carrier wave is known as a modulator. The process of modulating a carrier wave is a mixing process, and so the modulated wave is actually a combination of two waves.

## FREQUENCY MODULATION

In frequency modulation, as in amplitude modulation, we start with a carrier whose frequency and amplitude are constant. Figure 2-4 shows the effect of using an audio voltage to frequency modulate a carrier wave. Here we can see the basic differences between AM and FM. Unlike AM, the FM modulating process does not affect the amplitude of the carrier wave. However, the frequency of the carrier wave does change with the application of the audio modulating voltage. The change in frequency of the carrier is greatest when the audio modulating voltage is strongest. Thus, the strength of the audio signal determines the amount of frequency deviation of the carrier (Fig. 2-5).

What about the frequency of the audio modulating voltage? It will determine the rate at which the modulating changes will take place, that is, the frequency of the audio voltage will set the speed at which frequency changes take place.

## ADVANTAGES AND DISADVANTAGES

Amplitude modulation and frequency modulation (Fig. 2-6) each have

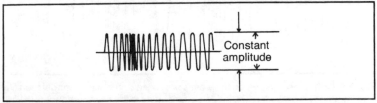

Fig. 2-4. FM wave has constant amplitude but frequency varies.

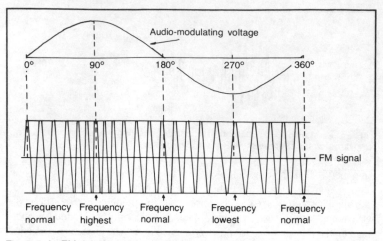

Fig. 2-5. In FM the change in carrier frequency depends on the strength, from moment to moment, of the modulating signal.

their advantages, and, of course, their disadvantages. The advantage of AM is that it requires a lesser amount of space; its bandwidth requirements are not large. Frequency modulation, on the other hand, has a much greater bandwidth requirement, but is less prone to signal degradation by electrical noise. Electrical noise behaves very much like a radio signal and can be produced by electrical machinery, vehicles, planes, fluorescent lights, electronic equipment, or static due to weather conditions. In short, electrical noise behaves as though it were an amplitude modulated signal. For this reason, electrical noise signals that can play havoc with AM broadcasts, are much less of a problem with FM (Fig. 2-7).

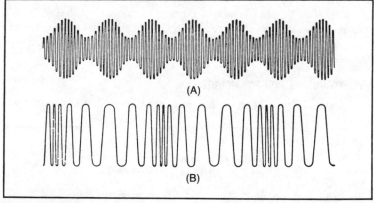

Fig. 2-6. Comparison between the waves produced by amplitude modulation (A) and frequency modulation (B). In AM the amplitude of the carrier changes; in FM, the amplitude of the carrier remains constant but the frequency changes.

46

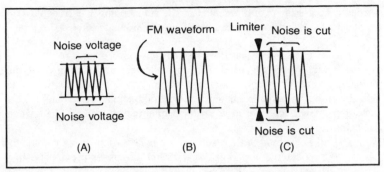

Fig. 2-7. Frequency modulated satellite signal audio and video carriers have constant amplitude, but noise voltages amplitude modulate the carrier (A). The signal plus added noise voltages following amplification (B). Limiter circuits cut noise voltages, restoring FM signal to its constant amplitude form (C).

## MODULATION OF THE TERRESTRIAL TV SIGNAL

Terrestrial television signals, whether VHF or UHF make use of both forms of modulation, AM and FM. The audio signal is frequency modulated; the video signal is amplitude modulated. Frequency modulation is not used for the picture signal, since its bandwidth requirements are far too great.

Now consider the uplink signals. These consist of audio and video, but unlike terrestrial TV signals, both are frequency modulated. Not only are the uplink signals frequency modulated, but the downlink signals are frequency modulated as well. The transponders change the carrier frequency , but continue to use the same form of modulation—that is, FM for both audio and video signals.

Television receivers are designed to accommodate terrestrial TV signals, not satellite TV. This means television receiver circuitry is made for amplitude modulated video and frequency modulated audio.

## SATELLITE CHANNEL BANDWIDTH

The total bandwidth allotted to the uplink signals is 500 MHz. Each uplink television channel is permitted 40 MHz. This may not seem like much but consider that the channel width of an earthbound television signal is 6 MHz, and so that used by an uplink channel is six times as much. This is necessitated by the fact that the uplink signals are frequency modulated.

The total bandwidth available to all the transponders in any of the satellites is 500 MHz. Since the permitted space allotted to each video program is 40 MHz it is apparent that the maximum number of channels available is 500 divided by 40 or 12, with a remainder of 20 MHz.

## NUMBER OF TRANSPONDERS

Since there are 12 available channels available in the uplink band, this determines the number of transponders on board a satellite. However, through the use of several techniques, the actual number of transponders can be doubled, and so most satellites have 24, although some do have only 12.

The first method is that the carrier frequencies of each of the transponder's carriers are staggered. Thus, as indicated earlier in Fig. 1-19, the frequency of every even numbered channel is selected so that it is positioned at the outer frequency limits of odd numbered channels. This arrangement was chosen so as to have minimum interference between channels.

The other method is the use of cross polarization, more commonly known as vertical and horizontal polarization, discussed previously. In a satellite using 24 transponders, no two adjacent channels have the same polarization. Thus, the carrier of a horizontally polarized channel is positioned between a pair of channels that are vertically polarized.

## GUARD BANDS

Another technique for keeping adjacent channels from interfering with each other is through the use of guard bands (Fig. 2-8). Thus, the actual TV signal bandwidth is kept to 36 MHz with a guard band of 4 MHz between channels.

While each channel contributes a total of 4 MHz for guard band use, this air space donated by each channel is arranged in two sections. Thus, Channel 2, for example, supplies 2 MHz at its lower frequency end and another 2 MHz at its upper frequency end, hence Channel 2 has a pair of buffer zones, each 2 MHz wide, for a total of 4 MHz. This same arrangement is applicable to all the other channels.

From all of this it would seem that a single satellite with 24 transponders would use up all the available channel space, and so it does. This same channel space is available to all the other satellites as well. There is no interference since the dish used for picking up the satellites may be a type that points at just one satellite at a time.

With a total available bandwidth of 500 MHz and the 480 MHz used by the transponders, we have a surplus of 20 MHz. This doesn't sound like much, but consider that this is still about 20 times as wide as the entire AM broadcast band. Rather than have this available bandwidth go to waste, it is

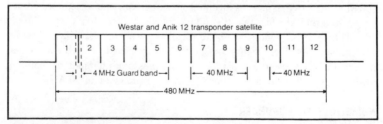

Fig. 2-8. 4 MHz guard bands are used between channels to prevent adjacent channel interference. (Radio-Electronics)

used for ground-to-satellite command signals, satellite-to-ground acknowl-edgement signals, plus a number of radio beacons. Radio beacons perform navigational functions and in this instance the satellite's radio beacons let ground control measure the exact position of the satellite.

## STATION IDENTIFICATION

There are a number of ways of identifying a station. On the AM broadcast band, a station is generally listed on the basis of its carrier frequency and the call letters of the station. The same approach is used for stations on the FM broadcast band. For terrestrial television stations, the carrier frequency isn't mentioned. Instead, the station is identified by its channel number. As an example, Channel 2 (as do all the other TV channels) has a total channel width of 6 MHz, covering the range from 54 MHz to 60 MHz. Since sound signal must accompany the picture signal, there are two carriers, not just one, as in the case of AM and FM broadcasting. For Channel 2, the video carrier is 55.25 MHz and the sound carrier is 59.75 MHz.

Satellite transponders, corresponding to terrestrial TV stations, are also identified by channel number. Since all the satellites can use the same channels, further identification is needed and this is done either by using the satellite's name or its alphanumeric designation, a letter followed by a number, as indicated in Fig. 1-13.

## MULTIPLE CARRIERS

AM or FM broadcast radio stations use just a single carrier, for they transmit just a single signal—audio. A television station, though, is in-volved in the radiation of two or more signals, depending on whether the transmission is monochrome or color.

The drawing in Fig. 2-9 illustrates the use of two carriers for a television signal. One of the carriers is for the picture; the other is for the sound. While these two carriers are independent of each other, they bear a fixed relationship to each other in terms of frequency. In the drawing, the two carriers are represented by vertical lines and are separated by 4.5 MHz. The drawing, however, does not indicate the actual carrier frequen-cies, simply their relationship. Thus, the video carrier for terrestrial TV station Channel 3 is 61.25 MHz, while its sound carrier is 65.75 MHz. The video carrier for Channel 4 is 67.25 MHz, while the sound carrier is 71.75 MHz. Note that the frequency difference between these two carriers, as it is for all TV channels, is 4.5 MHz.

Since the two carriers are independent, they can be modulated in different ways. As mentioned earlier, the video carrier is amplitude mod-ulated; the sound carrier is frequency modulated.

## THE SUBCARRIER

A further examination of Fig. 2-9 shows the presence of still another

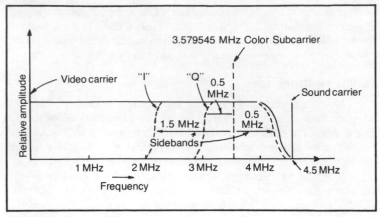

Fig. 2-9. The composite video signal has three carriers: a video carrier, a color subcarrier, and a sound carrier.

carrier positioned at 3.58 MHz above the picture carrier. This carrier is used for carrying color information. It is called a subcarrier because color is considered as part of the picture signal. The fact that it is called a subcarrier doesn't mean it is inferior in any way, for it is still an independent carrier. Like the picture carrier, the color subcarrier uses amplitude modulation.

## TRANSPONDER SUBCARRIERS

Like earthbound television stations, the transponders used in satellites also employ subcarriers for the transmission of additional signals. One of these subcarriers is for the audio signal, and, like terrestrial TV signals, the audio and picture carriers are separated by a fixed frequency amount. For terrestrial stations, the frequency separation is 4.5 MHz. For transponder transmissions, the frequency separation can be 5.6 MHz, 5.8 MHz, 6.2 MHz, or 6.8 MHz. Since separate subcarriers are used for audio information, the satellite receiver must be equipped with what is in effect an audio tuning control.

## THE AUDIO ONLY TRANSPONDER

While a TVRO is primarily concerned with the reception of picture signals, a transponder could be used mainly for the transmission of audio. A single transponder can handle dozens of audio signals and can do so in several ways. Since the required bandwidth of an audio signal is far narrower than that of a video signal, measured in kHz rather than in MHz, and since the available bandwidth per channel is 40 MHz, a single transponder is capable of carrying hundreds of voice broadcasts and music programs. It is also possible for a transponder to make use of a technique known as single channel per carrier, or SCPC. Transponders handling audio only signals sometimes use much higher power for transmission than video. As a result, the dishes made for sound only reception can be made smaller than those used for video signal pickup.

## MONOPHONIC AND STEREOPHONIC SOUND

For satellite transmission of monophonic sound, just a single subcarrier is needed. If a satellite is equipped with 24 transponders, for example, the same subcarrier frequency can be used for a particular audio purpose, such as mono sound accompanying video.

For stereophonic sound, a pair of subcarriers are used, one for each sound channel but these stereo sound subcarriers need not be the same for each program. Thus, Satcom F3 uses one pair of subcarriers (5.40 and 5.44 MHz) for one program and 5.50 MHz and 5.76 MHz for another program. When a single subcarrier is used, it is sometimes referred to as subcarrier A. When a pair of subcarriers are used, for stereo transmission, they are identified as A and B, or A + B.

Not only can a satellite use one or a pair of sound subcarriers, but also more than one pair. The frequency separation of these subcarriers depends on the satellite. Thus, Satcom F3 uses a pair of subcarriers of 5.40 and 5.94 MHz, with a frequency separation of 0.54 MHz between them (5.94 − 5.40 MHz = 0.54 MHz). The same satellite uses another pair of subcarriers, 5.50 and 5.76 MHz, with a frequency separation of 0.18 MHz (5.76 − 5.58 = 0.18).

There are two types of stereo: matrix and discrete. With discrete stereo, there will be two active subcarriers with each having the same amount of signal level. One of the subcarriers transports the left audio signals; the other carries the right audio signals. Each of these subcarriers must be tuned to receive a discrete stereo transmission.

With matrix stereo, as with discrete, two active subcarriers must be received. One of the subcarriers will be substantially louder than the other. The lower level audio is known as the left minus right subcarrier, also known as the L − R or difference subcarrier. The louder audio signal is called the left plus right (L + R) or audio subcarrier sum frequency.

## DATA SUBCARRIER

It is also possible for a transponder to have not just one or two subcarriers, but a number of them. These subcarriers can be used to supply the viewer with data or radio broadcast programs. Since these are on separate subcarriers, they do not interfere with the reception of the video signal, nor can they be heard or otherwise utilized unless the TVROs satellite signal receiver is equipped to do so.

## BASEBAND SIGNALS

The audio and video signals that are modulated onto carrier waves are known as baseband signals. The video baseband signal, more often simply called the video signal, has a frequency range from about 50 Hz to 4.2 MHz.

51

The audio baseband signal extends from about 30 Hz to about 15 kHz. These two signals, plus various pulses for controlling the movement of the electron scanning beam in the television picture tube, are referred to as the composite video signal.

## DEMODULATION

Very high frequency carriers are used to deliver the audio and video signals from earth to a satellite and also from the satellite to a receiving station on earth. As long as the carrier is to be used in space, its frequency is that specified by the FCC. Once the signal has been received, it travels through components and cables. In its passage through these, the carrier frequency is lowered, but since the carrier is confined to equipment and wires, it can have any frequency selected by the manufacturer.

To put the baseband signals on a new carrier, the signal must first be demodulated. This means that the original baseband signals, video and audio, are recovered. They are then used to modulate a lower frequency carrier. This process may be repeated several times, and so ultimately the carrier frequency is far below that used for space transportation.

# Chapter 3

# The TVRO System

You can mix and match satellite TVRO components: dishes, LNAs, downconverters, receivers, rotators, and remote control units from different manufacturers. Some satellite TV enthusiasts follow this method for budgetary reasons or with the expectation that they can build a better overall system this way.

This approach has its hazards, for the selected components may not work well together. They may not interface properly or as well as they should. It also calls for considerable expertise on the part of the buyer since no satellite system can perform any better than its poorest component. To hook up an excellent LNA with a less than desirable dish will not result in signal improvement. A safe procedure is to select products of an established, reputable manufacturer and to use all components of the same brand name.

## WHO OWNS THE SATELLITES?

Launching a satellite is expensive, but is only part of the overall cost, measured in millions of dollars, with sums such as $10 million to $15 million not uncommon. This does not include the cost of designing and building the satellite, maintaining it in its proper orbit and supplying it with programs for re-transmission to earth.

Some companies do not use all the transponders on their satellites and rent some of these to companies who either cannot afford satellite costs or who have a need for only one or two transponders. Thus, a company specializing in the production of video programs could use one of the satellite's transponders for sending programs to a large group of cable companies. In some instances, the transponders are rented; in others, they are purchased outright.

## EARTH STATION

The words *earth station* are used to describe any station on earth that is used not only for the reception of satellite signals but for their transmission as well, that is, a station involved in handling both uplink and downlink signals. A cable company, for example, could be an earth station operator, or an earth station could be a company that develops video programs.

## TELEVISION RECEIVE ONLY

A TVRO is quite different from an earth station, for TVROs are only involved in signal reception. TVROs do not require station licensing; earth stations do.

There are three basic types of TVROs, but while these are placed in different categories, the equipment used by all three is essentially the same.

The first category includes satellite installations used by the armed forces and various commercial enterprises. Unlike home-type TVROs, these are used to pick up weather data, security information, and streams of business data.

The second category includes cable companies, motels, hotels, and apartment complexes. In this second category, the TVRO system may be used as the program source for any number of television receivers.

The third category includes TVROs for the reception of satellite TV signals in the home. This category, possibly the smallest at this time, will probably be the largest in the future. Its rate of growth is more than that of the other two categories combined.

## THE TVRO SYSTEM

The basic structure of the TVRO system was mentioned in Chapter 1 but we can now begin to expand on it. The television signal is transmitted from an earth station to a selected satellite, but before transmission, the signal requires processing, with the amount and type of signal processing depending on the original signal source.

The processing used at the earth station prior to uplink signal transmission depends on the condition of the signal. If the signal is directly baseband, such as that supplied by a video tape or video disc, both the audio and video portions of the signal must be frequency modulated using a radio-frequency carrier in the uplink portion of the C band, that is, a carrier having a frequency in the range of 5.925 to 6.425 GHz.

This does not mean that just a single uplink carrier will be used. One will be required for the video waveform, and one or more will be needed for the audio, depending on whether it is mono or stereo.

It is also possible that the program to be uplinked will be delivered to the earth station in the form of an earthbound television signal. In that case the signals will already have been modulated, using AM for the video, FM for the audio, onto a carrier having a frequency in the VHF or UHF bands. For signals such as these, it is first necessary to demodulate—that is, to remove the carrier and restore the audio and video signals to baseband form. After this is done, the signals can then be used to frequency modulate an uplink carrier wave.

## PRE-EMPHASIS

Pre-emphasis, as its name implies, is a technique for emphasizing part of a signal prior to transmission. In the case of earthbound FM stations, for example, pre-emphasis is used to improve the signal-to-noise ratio. Electrical noise seems to have an affection for the treble range of an audio signal, and further, the energy contained in the treble range is lower than that in the bass and midrange tones. Pre-emphasis, then, is used to give the treble range a boost, comparable to turning up a treble gain control. In the receiver, a de-emphasis circuit is used to counter the action of pre-emphasis.

Prior to the transmission of the uplink signal, both the audio and video signals are pre-emphasized. This action supplies a boost to the high frequency portion of the signals, and, just as in the case of pre-emphasis for earthbound FM signals, is used to improve the signal-to-noise ratio.

Another reason for both audio and video pre-emphasis is that demodulators, circuits for removing the baseband signals from their carriers, become noisier with increases in frequency. Pre-emphasis helps compensate for that demodulator operating characteristic. In the satellite receiver, pre-emphasis is countered by including a de-emphasis circuit following the demodulator. This circuit has no panel controls and so its functioning is completely automatic.

## DITHERING

In addition to pre-emphasis, another signal processing technique is used, known as dithering. Dithering is a form of signal energy dispersal and is applied to the video baseband signal prior to uplink modulation.

In itself, dithering is a form of modulation, but instead of a waveform such as that used by a carrier, employs a waveform that looks like a triangle.

The composition of a video signal consists of two interlaced fields, each of $262\frac{1}{2}$ lines, forming a single picture frame (Fig. 3-1). What this means is that a television picture is *drawn* on the inner face of the television picture tube, a line at a time. The first $262\frac{1}{2}$ lines are traced and these lines are called a field. Then the next $262\frac{1}{2}$ lines are drawn and are also called a field. The lines of each field alternate. Two fields constitute a complete frame and since each field has a repetition rate of 60 per second, a frame or complete picture has a rate of 30 per second.

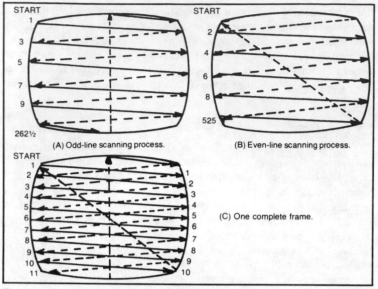

Fig. 3-1. Odd line scanning (A) and even line scanning (B). Each scanning consists of 262½ lines, called a field. The two fields form a single complete picture, or one frame (C).

The dithering frequency is also at a 30 per second rate. The purpose of dithering is to spread the picture signal energy more uniformly over its band.

Dithering has several advantages. Because the signal energy in the baseband video signal is no longer lumped, that is concentrated at certain frequencies, there is less chance of interference with adjacent channels. An additional benefit is that it reduces the possibility of interference with other services, such as earthbound microwave links used by telephone companies.

## VIDEO CLAMPING

Ultimately, the baseband signals will need to be presented at the antenna input terminals of the in-home television receiver. Prior to that point the various processing steps will need to be undone. Pre-emphasis will need to be de-emphasized. Modulation will need to be demodulated and dithering will need to be *undithered*. The circuit that will do the so-called undithering will be a video clamping arrangement. Without this, the result would be picture flicker at a 30 Hz rate.

Various processing steps are taken prior to transmission of the downlink signal. A number of audio subcarriers can be used. Typical subcarrier frequencies for the audio signals are 5.6, 5.8, 6.0, 6.2, and 6.4 MHz. Note that the frequency separation for these frequencies is 0.2 MHz or 200 kHz. This follows the same procedure as that used for earthbound

FM in which each FM channel is allotted 200 kHz. The maximum permitted modulation is 150 kHz, with a 25 kHz guard band on either side. The purpose of the guard band is to minimize interference with adjacent FM channels.

## UPCONVERSION

The C band is actually two bands of frequencies. One is used for uplink signals and, as indicated earlier, has a bandwidth of 5.925 to 6.425 GHz. The other is used for the retransmission of signals from the satellite's transponders and its bandwidth is from 3.7 to 4.2 GHz.

The baseband signals must then be loaded onto a very high frequency carrier wave in the uplink portion of the C band. At the earth station the modulation process is carried on in several steps. The first is to establish a carrier frequency of 70 MHz, with the baseband signals modulated onto this carrier. The video signal, following modulation, has a maximum bandpass of 36 MHz, but various audio subcarriers must be accommodated as well.

Sometimes the 70 MHz amplifier following the modulator is referred to as an intermediate frequency amplifier. This amplifier has a double function. Its purpose is to amplify the modulated signal and also to work as a filter, removing any interfering signals outside the 70 MHz band. The concept of an intermediate frequency (i-f) amplifier will be discussed in more detail in connection with receivers.

The next stage in the earth station transmitter is an upconverter circuit. In this circuit, the 70 MHz carrier is eliminated and a carrier of the desired uplink frequency is substituted. The signal is then amplified once again in a power amplifier. It is the power amplifier that supplies the signal with its electrical energy, generally in the order of a half kilowatt. This isn't very much if you consider that the average large screen television set is about 300 watts or that an electric heater may require 1,200 watts. A half kilowatt, or 500 watts, is the amount of energy needed by five 100 watt electric light bulbs. Not very much.

## SATELLITE POWER

Satellites are comparatively small stations and the physically largest component is the solar panel. If the satellite is equipped with a pair of dishes, one could be used for signal reception; the other for signal transmission. Since the satellite may have as many as 24 transponders, the same pair of dishes can be used for each channel. This is done by dish sharing, a switching method in which each transponder is given an opportunity to receive and transmit signals for a brief time, actually small fractions of a second.

The composite video signals transmitted to earth via each transponder frequency channel thus does not consist of a continuous signal but one which is transmitted in segments, with the time interval between segments extremely short.

To the viewer watching TV on earth, the picture appears continuous, unbroken. Because of a characteristic of the human eye known as persistence of vision it isn't possible to detect an absence of picture during no picture time.

Each transponder in the satellite has its own operating band of frequencies, just as all the television stations on earth have their own. Because of this frequency difference each individual transponder of a satellite can be tuned in by a receiver on earth. A single satellite with 24 transponders can supply 24 channels of video programming, twice as many as the 12 channels available to earth TV stations. That is for just one satellite. The next satellite, equipped with 24 transponders could also supply 24 television channels. The programs of this satellite could be received by a small readjustment of the position of the receiving dish on earth.

If a dish on earth has a clear sight to all the satellites, it can receive all the channels of those satellites. Since more satellites are being placed in orbit, the total program capability of satellites could be measured in the hundreds, far exceeding the combined capabilities of VHF and UHF earthbound channels.

## HOW MANY CHANNELS?

It would be easy enough to assume that all satellites have 24 transponders and so, multiplying this by the total number of satellites could readily produce an answer in the hundreds. There are several factors that reduce this number considerably.

The first of these is that not all dishes have a clear view to all satellites. Because of hilly terrain, buildings, or trees it may be possible to have a clear view of just a few satellites. Still another factor is that even if a satellite does have 24 transponders this is no indication that these are all working at the same time. Some may be used for services other than video. Finally, not all satellites are equipped with 24 transponders. Some have only 12. Some satellites to be launched on the Ku band will have less, possibly 3 to 6. Even with these limitations, satellites have a far greater program capability than terrestrial television stations.

## THE TRANSPONDER

The purpose of a transponder is to receive the signal from the earth station and then to relay it to earth. The transponder is not a signal reflector, but rather a combined receiver/transmitter. The input to the transponder is a 500 MHz filter. This is a wide bandpass device designed to cut off all frequencies below the C band uplink and all frequencies above it. Its purpose is to make sure the receiver section of the transponder responds only to uplink signals.

Following the input filter, we have a receiver. This receiver is a superheterodyne using circuitry similar to that of the earthbound satellite receiver in the home. The receiver contains a frequency converter, a mixer circuit that also includes a local oscillator. The signal generated by the local oscillator circuit mixes or heterodynes the incoming uplink signal and the result of this action is a lower frequency signal, a process known as signal downconversion.

The downconversion procedure does not affect the video or audio baseband signals. In effect, what it does is to remove the carrier used to transport the signal from earth to the transponder. Having disposed of this carrier, it supplies a new one and this new carrier is the frequency of the downlink signal. Downconversion, then, is the process of substituting a lower frequency carrier for one that has a higher frequency.

Since the downconversion process not only results in a lower frequency carrier, but a multiplicity of other signals as well, a 36 MHz bandpass filter is used to eliminate them. The result is a downlink signal having a bandpass of 36 MHz. This signal is then strengthened in a power amplifier and fed to a transmitting antenna using a dish as a reflector.

## MICROWAVE REPEATERS

A transponder is also known as a microwave repeater, a carryover from the use of this term for earthbound stations used for information transmission overland. Basically, the idea is the same. The microwave repeater receives a signal, amplifies it, and sends it on its way to the next repeater station, and so on until it reaches its final destination.

## THE HOME TELEVISION STATION

TVRO, or television receive only, implies that the station isn't used for two-way communications. It does not mean the TVRO consists only of one television receiver, for the signal received from a satellite can be used to supply one or more TV sets in the same home, or TV receivers in different homes, possibly adjoining, or the TV sets in each apartment of an apartment house, or separate accommodations in a motel or hotel. Whether the signal is delivered to just one TV set or to many, just a single dish is used.

## THE DISH

The dish is simply a device for picking up as much of the satellite TV signal as possible, focusing that signal on a central point somewhat forward of the dish itself.

The dish is a passive device. It does not amplify the signal, nor does it tune it in. All that the dish can do is to work as a signal collector and the larger the diameter of the dish the greater its signal collecting ability.

Not only must the dish collect the signal but it must focus it accurately and this is a function of the shape of the dish. The focusing action is quite critical, for if the dish cannot focus the signal correctly the result can be a poor picture or no picture at all. A more nearly appropriate name for a dish would be *signal reflector*.

## THE FEEDHORN

The picture signal is maximum at the focal point of the dish, for it is concentrated here by reflection from the surface of the dish. At this focal point we have a component called a feed. The feed is simply the entrance to a short length of metallic tubing known as a waveguide.

At the entrance to the waveguide there may be a device called a scalar feed. This consists of a number of concentric rings that correspond to the circular shape of the dish. These rings also function as reflectors, helping to direct the outer part of the focused signal toward the center of the feed. Consider the feedhorn as a receptor positioned at the focal point of the dish.

Not all of the dish is equally desirable for signal reflection. The most useful portion can be considered as that area extending from the center of the dish to about 70% of the way toward the outer edge. The design of the feed is such that it accepts signals from this portion of the dish but attenuates the remaining part. It is this section of the dish that picks up unwanted electrical earth generated noise.

Optimum signal pickup is desirable but if the noise level is too high, that is, if the signal-to-noise ratio is poor, then it is better to reject part of the signal if it also means rejecting noise.

## THE WAVEGUIDE

Signals can travel in three ways. A terrestrial TV station transmits a composite video signal superimposed on a carrier wave. Such waves, whether sent through space from the antenna of a terrestrial station or the transponder in a satellite are unbounded, that is, they aren't confined in any way. The tendency of such waves after leaving the antenna is to spread out. The wave energy becomes diffused.

Another method by which radio waves travel is through transmission lines. A TV antenna is actually a probe set up in the energy field of the radio wave. It picks up a small amount of signal, and, using the conducting wires of either two-wire transmission line or coaxial cable, delivers the signal energy to the antenna input terminals of a TV set.

The third is the enclosed method, with the enclosure known as a waveguide. A waveguide can be a circular tube or one that is rectangular. Both types are hollow, and can be used together. Waveguide can be flexible or non-flexible, with the latter type more commonly used. The entire waveguide is a feed delivering satellite signals to an antenna. However, the entrance to the waveguide is often called the feed.

There is a relationship between the dimensions of a waveguide and the frequency of the waves it conducts, so waveguides are suitable only for the microwave frequencies used in satellite TV.

There is nothing mysterious or magical about waveguides. They are just a form of transmission line and so are related to twin lead and coaxial cable. The length of the waveguide used in satellite signal reception is quite short and is measured in inches.

Waveguide is used to conduct signals from the focal point of the dish to the input of the following component, a low-noise amplifier (LNA).

## THE ANTENNA PROBE

Although the dish is often referred to as an antenna, it is not. To pick up the television signal, a tiny probe is inserted into the end of the waveguide. This probe, which can have a variety of shapes, is the actual antenna. Coaxial cable is used to connect the antenna probe to the input of the LNA. In some instances, the antenna is just a continuation of the hot lead of the coaxial cable. The metallic shield of the cable is connected to the metal shell of the waveguide and in this example the hot lead of the coaxial cable is the antenna probe. The probe can also be in the form of a U or can be circular.

Antenna size is a function of frequency, with the antenna regarded as a broadly tuned circuit. In this case, the length of the antenna probe is such that it is resonant to the approximate center of the downlink C band. As television frequencies become higher, the antenna pickup becomes smaller. Thus, FM antennas are smaller than antennas cut to pick up earthbound Channels 2 through 5. UHF antennas, since they must respond to much higher frequencies, are smaller than FM antennas. Because satellite TV signals are in the microwave region, the antenna pickup is smaller still.

The probe antenna can be rotated so as to pick up either vertically polarized or horizontally polarized satellite signals. The turning device, a servomotor, is remotely controlled from the home. In some instances the entire feed assembly is rotated but with the antenna probe remaining fixed in position. A preferable arrangement is to have the antenna probe and the LNA remain fixed in position, but the energy fields of the signal rotated.

## THE LNA

The LNA accepts the signal from the feed via a very short length of coaxial cable. The LNA is a broadband amplifier having extremely high gain, with as low a noise level as possible. The LNA doesn't alter the frequency of the signals it receives from the waveguide, but simply strengthens them. Since the frequencies involved are so high, the input and output connections to the LNA must be as short as possible. For this reason the LNA is often made a part of the waveguide assembly.

## THE DOWNCONVERTER

The amplified signal supplied by the LNA could be brought into the home via coaxial cable for further processing. However, at the 4 GHz output of the LNA, the frequency of the carrier wave is so high that it would result in substantial signal loss. Special types of coaxial cable could be used (and were used at one time) but such cable, known as Heliaz, is quite expensive. The obvious solution is to use a downconverter positioned directly at the LNA output. This means the LNA and its immediately following downconverter are made part of the waveguide assembly and are located outdoors. As a consequence, both the LNA and its downconverter are housed in a waterproof assembly. Neither the LNA nor the downconverter have any operating controls and once fixed in position generally require no further attention.

All radio receivers, whether AM or FM broadcast, use a superheterodyne circuit, an arrangement applied to the downconverter, as indicated in Fig. 3-2. The unit is equipped with a local oscillator, a tunable electronic generator. The signal produced by the local oscillator is brought into a mixer circuit, and the two signals, the incoming signal and that produced by the local oscillator are *mixed* or heterodyned. The result is a number of frequencies, but the most notable of these are the two original frequencies, and another pair, one of which is the sum of the two original frequencies and the other is their difference. The difference frequency, the local oscillator frequency minus the input signal frequency, could be selected. Known as an intermediate frequency, abbreviated as i-f, it generally has a value of 70 MHz.

If, for example, the incoming signal has a frequency of 3.7 GHz, then the local oscillator has a frequency of 3,770 MHz. 3,770 MHz − 3,700 MHz = 70 MHz. If the incoming frequency is higher (it can go as high as 4.2 GHz) then the local oscillator is tuned so it is always 70 MHz higher in frequency. Thus, the intermediate frequency always remains constant at 70 MHz, no matter what transponder is selected.

The local oscillator is tuned by the application of a dc voltage, hence is known as a voltage controlled oscillator or VCO. It is also referred to as a voltage tuned oscillator or VTO.

## THE SATELLITE RECEIVER

From the downconverter, the signal is delivered to a satellite receiver

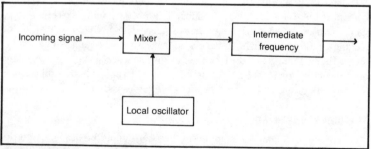

Fig. 3-2. Downconversion is based on the superheterodyne principle. The incoming signal is beat (mixed or heterodyned) in a mixer circuit. The output is a new, much lower frequency known as the intermediate frequency (i-f).

62

located in the home. This receiver also uses the principle of superheterodynes, and may be equipped with a local oscillator/mixer or downconverter circuit. The result of this downconversion process is an intermediate frequency signal.

Following the i-f stages in the receiver, the carrier, having completed all its functions, is disposed of. This function is done by a demodulator circuit. What we have, then, at the output of the demodulator, are our original baseband signals, both video and audio. As they are, they cannot be brought into a television set, for these receivers cannot accommodate baseband signals without an accompanying carrier.

Since this is a requirement of the TV set it may seem that we disposed of the carrier a bit too soon. There are two factors that made that carrier useless for our purpose. The first is that the carrier was frequency modulated for both video and audio. TV sets are designed to accommodate only signals that have amplitude modulation for the video and frequency modulation for the audio. Consequently the form of modulation we have been using up to this point is of no value.

The other factor is the frequency of the carrier. For a video and audio signal to be suitable for a TV set, the frequency of its carrier must be the same as the frequency of any terrestrial TV station. This frequency could be any one of those used for Channel 2 through Channel 5, or Channel 6 through Channel 13, or any of the UHF channel frequencies. Satellite receivers generally modulate the recovered baseband signals with a carrier having the frequency used by Channel 3 or Channel 4, both VHF frequencies. Consequently, the output of the satellite receiver can be connected directly to the antenna input terminals of the television receiver. That receiver must have its channel selector set to either Channel 3 or Channel 4 reception, depending on which of these carriers is being used.

We have now had two overviews of the TVRO system, but these, even taken together, are no more than a bare introduction. In the chapters that follow we can examine each of the TVRO components in more detail.

## SPECIFICATIONS

You can become familiar with a manufacturer's product line by visiting his dealers' showrooms, by reading his advertising, or by reading the available literature. Various magazines also describe TVRO products and some of them report on tests they make. Another method is to read specification sheets (specs) issued by manufacturers. These represent the results of tests performed by manufacturers' laboratories. The advantage of a spec sheet is that you can easily make a comparison of products, provided you understand the language of the spec sheet. These are explained for each of the components in the chapters that follow.

## DECIBELS

The purpose of a spec sheet is to make a product evaluation and this is

| Voltage or current ratio | Power ratio | − dB + | Voltage or current ratio | Power ratio |
|---|---|---|---|---|
| 1.000 | 1.000 | 0 | 1.0000 | 1.0000 |
| 0.989 | 0.977 | 0.1 | 1.0116 | 1.0233 |
| 0.977 | 0.955 | 0.2 | 1.0233 | 1.0471 |
| 0.966 | 0.933 | 0.3 | 1.0351 | 1.0715 |
| 0.955 | 0.912 | 0.4 | 1.0471 | 1.0965 |
| 0.944 | 0.891 | 0.5 | 1.0593 | 1.1220 |
| 0.933 | 0.871 | 0.6 | 1.0715 | 1.1482 |
| 0.923 | 0.851 | 0.7 | 1.0839 | 1.1749 |
| 0.912 | 0.832 | 0.8 | 1.0965 | 1.2023 |
| 0.902 | 0.813 | 0.9 | 1.1092 | 1.2303 |
| 0.891 | 0.794 | 1.0 | 1.1220 | 1.2589 |
| 0.881 | 0.776 | 1.1 | 1.135 | 1.288 |
| 0.871 | 0.759 | 1.2 | 1.1482 | 1.3183 |
| 0.861 | 0.741 | 1.3 | 1.161 | 1.349 |
| 0.851 | 0.724 | 1.4 | 1.175 | 1.380 |
| 0.841 | 0.708 | 1.5 | 1.189 | 1.413 |
| 0.832 | 0.692 | 1.6 | 1.202 | 1.445 |
| 0.822 | 0.676 | 1.7 | 1.216 | 1.479 |
| 0.813 | 0.661 | 1.8 | 1.230 | 1.514 |
| 0.803 | 0.646 | 1.9 | 1.245 | 1.549 |
| 0.749 | 0.631 | 2.0 | 1.2589 | 1.5849 |
| 0.776 | 0.603 | 2.2 | 1.288 | 1.660 |
| 0.759 | 0.575 | 2.4 | 1.318 | 1.738 |
| 0.750 | 0.562 | 2.5 | 1.334 | 1.778 |
| 0.724 | 0.525 | 2.8 | 1.380 | 1.905 |

| Voltage or current ratio | Power ratio | − dB + | Voltage or current ratio | Power ratio |
|---|---|---|---|---|
| 0.141 | 0.0200 | 17 | 7.08 | 50.1 |
| 0.126 | 0.0159 | 18 | 7.94 | 63.1 |
| 0.112 | 0.0126 | 19 | 8.91 | 79.4 |
| 0.10000 | 0.0100 | 20 | 10.00 | 100.0 |
| 0.08913 | 0.0079 | 21 | 11.22 | 125.9 |
| 0.07943 | 0.0063 | 22 | 12.59 | 158.5 |
| 0.07079 | 0.0050 | 23 | 14.13 | 199.5 |
| 0.06310 | 0.00398 | 24 | 15.85 | 251.2 |
| 0.05623 | 0.03162 | 25 | 17.78 | 316.2 |
| 0.05012 | 0.002512 | 26 | 19.95 | 398.1 |
| 0.04467 | 0.001995 | 27 | 22.39 | 501.2 |
| 0.03981 | 0.001585 | 28 | 25.12 | 631.0 |
| 0.03548 | 0.001259 | 29 | 28.18 | 794.3 |
| 0.03162 | 0.001000 | 30 | 31.62 | 1000 |
| 0.02818 | 0.000794 | 31 | 35.48 | 1259 |
| 0.02512 | 0.000631 | 32 | 39.81 | 1585 |
| 0.02239 | 0.000501 | 33 | 44.67 | 1995 |
| 0.01995 | 0.000398 | 34 | 50.12 | 2512 |
| 0.01778 | 0.000316 | 35 | 56.23 | 3162 |
| 0.01585 | 0.000251 | 36 | 63.10 | 3981 |
| 0.01413 | 0.000199 | 37 | 70.79 | 5012 |
| 0.01259 | 0.000158 | 38 | 79.43 | 6310 |
| 0.01122 | 0.000126 | 39 | 89.13 | 7943 |
| 0.01000 | 0.000100 | 40 | 100.00 | 10000 |
| 0.00891 | 0.000079 | 41 | 112.2 | 12590 |

| | | dB | | |
|---|---|---|---|---|
| 0.708 | 0.501 | 3.0 | 1.4125 | 1.9953 |
| 0.692 | 0.479 | 3.2 | 1.445 | 2.089 |
| 0.676 | 0.457 | 3.4 | 1.479 | 2.188 |
| 0.668 | 0.447 | 3.5 | 1.4962 | 2.2387 |
| 0.661 | 0.436 | 3.6 | 1.514 | 2.291 |
| 0.646 | 0.417 | 3.8 | 1.549 | 2.399 |
| 0.631 | 0.398 | 4.0 | 1.5849 | 2.5119 |
| 0.596 | 0.355 | 4.5 | 1.6788 | 2.8184 |
| 0.562 | 0.316 | 5.0 | 1.7783 | 3.1623 |
| 0.531 | 0.282 | 5.5 | 1.8836 | 3.5481 |
| 0.501 | 0.251 | 6.0 | 1.9953 | 3.9811 |
| 0.473 | 0.224 | 6.5 | 2.113 | 4.467 |
| 0.447 | 0.200 | 7.0 | 2.239 | 5.012 |
| 0.422 | 0.178 | 7.5 | 2.371 | 5.623 |
| 0.398 | 0.159 | 8.0 | 2.512 | 6.310 |
| 0.376 | 0.141 | 8.5 | 2.661 | 7.079 |
| 0.355 | 0.126 | 9.0 | 2.818 | 7.943 |
| 0.335 | 0.112 | 9.5 | 2.985 | 8.913 |
| 0.316 | 0.100 | 10 | 3.162 | 10.00 |
| 0.282 | 0.0794 | 11 | 3.55 | 12.6 |
| 0.251 | 0.0631 | 12 | 3.98 | 15.9 |
| 0.224 | 0.0501 | 13 | 4.47 | 20.0 |
| 0.200 | 0.0398 | 14 | 5.01 | 25.1 |
| 0.178 | 0.0316 | 15 | 5.62 | 31.6 |
| 0.159 | 0.0251 | 16 | 6.31 | 39.8 |
| 0.00794 | 0.000063 | 42 | 125.9 | 15850 |
| 0.00708 | 0.000050 | 43 | 141.3 | 19950 |
| 0.00631 | 0.000040 | 44 | 158.5 | 25120 |
| 0.00562 | 0.000032 | 45 | 177.8 | 31620 |
| 0.00501 | 0.000025 | 46 | 199.5 | 39810 |
| 0.00447 | 0.000020 | 47 | 223.9 | 50120 |
| 0.00398 | 0.000016 | 48 | 251.2 | 63100 |
| 0.00355 | 0.000013 | 49 | 281.8 | 79430 |
| 0.00316 | 0.000010 | 50 | 316.2 | 100000 |

Fig. 3-3. Voltage, current, and power ratios and their corresponding values in decibels. (From the *Master Handbook of Electronic Tables & Formulas, 3rd edition*, by Martin Clifford, published by TAB Books)

most often done by specific tests. A spec sheet will list mechanical features of a product and also electronic capabilities. Measurements are often indicated in terms of decibels.

Abbreviated as dB, the decibel is a unit of measurement, but unlike measurements such as the inch, yard, centimeter, or meter, isn't linear. For all measurements, whether linear or not, there must always be a reference. A measurement is a comparison, or the ratio, of two points, one with respect to the other, or else a ratio of two electrical quantities.

In the case of linear measurements made with a ruler, the left side of the ruler is zero and all measurements are made with reference to that point. If you decide to measure from 1 on that ruler instead, all you have done is shifted the reference point.

A decibel is a measurement between a pair of voltages, currents, or powers. If two values are supplied, one serves as the reference, the other the amount of voltage, current, or power being measured.

If there is no reference, that is, if just one value of voltage, current, or power is available, an artificial reference can be supplied and then the measurement is made with respect to this reference. The reference is usually indicated. It is also sometimes inferred when it is quite clear just what the reference is.

In the case of an active device, such as an amplifier, the gain of the amplifier is often supplied in decibels, and is a comparison of the output signal compared to that supplied at the input. The input in this case is the reference. For passive devices, such as dishes, the reference can be a standard dish or some arbitrary reference based on noise level.

## DECIBEL CALCULATIONS

Decibels are used in connection with the gain or loss of signal in a component. Coaxial cable, a passive device, has a signal loss, depending on the structure of the cable and the frequency of the signal. An amplifier, an active device, is often measured in terms of signal gain, expressed in dB.

While decibels can be calculated through the use of formulas, an easier way is to consult a table, such as the one shown in Fig. 3-3. This table is a listing of decibels and their voltage, current, and power ratios.

In the table, each power value in dB can be obtained by multiplying the preceding ratio by 1.259. For example, 6 dB corresponds to a power ratio of 3.9811. The value of 7 dB can be obtained by multiplying the value of 6 dB by 1.259. With this method, given any value of decibels, we can obtain any other value, also in decibels, by using a little arithmetic.

Of course it isn't necessary to do any arithmetic if a table, such as the one in Fig. 3-3 supplies the answers, but note the table is limited and ends with a value of 50 dB.

Now look at the chart once again. In going from 0 dB to 10 dB we have a power ratio of 10. In effect, the first digit in the right-hand column, 1, is multiplied by 10. Similarly, in going from 10 dB to 20 dB, the power ratio has become 100. Now look at 50 dB. It corresponds to a ratio of 100,000.

This means that if an amplifier has a power gain of 50 dB, the power ratio is 100,000.

Decibels can also be used for making comparisons between a pair of voltages or currents, but most often spec sheets for TVRO products use power ratios, in decibels. Decibels are not used for comparing a voltage to a current, or current to power. The table in Fig. 3-3 shows the comparison between voltage ratios and decibels. According to this table, an increase in voltage from 0 dB to 10 dB means that the output voltage of a component being measured has 3.162 times the voltage present at the input. The voltage being measured is signal voltage and does not refer to the dc operating voltage being supplied to the component. The table in Fig. 3-3 can also be used for making a comparison between two currents.

## REFERENCES

When using decibels in making measurements of power gain or loss, any amount of power can be used as the reference. To indicate the reference that has been selected, a letter is used immediately following the abbreviation dB. Thus, dBw means decibels referenced to 1 watt. 10 dBw means a gain of 10 dB above 1 watt. If the watt is too large for use as a reference, a milliwatt, (a thousandth of a watt) can be used. The abbreviation then becomes dBm. For extremely small amounts of power, the femtowatt is used. A femtowatt is one quadrillionth of a watt and in exponential form is written as $10^{-15}$ watt, or 0.000000000000001 watt. dBi means decibels of gain relative to isotropic and used to indicate the gain of a dish. Isotropic means identical in all directions.

## NOISE FIGURE

Noise is an ever-present part of all signals, whether AM, FM, TV broadcast, or satellite TV. Noise in this context does not mean noise that is directly audible, but refers to noise in the form of a voltage. Noise can be produced naturally since nature is a constant producer of electrical noise, but it can also be generated by any electrical component, such as a receiver, amplifier, or television set.

The term *noise figure* refers to the noise produced by any active signal handling component and is an indication of the amount of noise it adds to the signal, with 0 dB as the reference. Obviously, active components having low-noise figures are preferable. Noise figures are directly additive. If one component has a noise figure of 5 dB, and delivers a signal to another component having a noise figure of 4 dB, the total noise figure, known as the system noise figure, is the sum of the two or 9 dB.

67

# Chapter 4

# The Dish

The amount of power required by the transponder for the transmission of signals to earth in the C band is about 5 watts, but this 5 watts is dc input power to the final amplifier, the power amplifier. How much of this five watts reaches a dish on earth depends on a number of factors, including the conversion efficiency of that final power amplifier, that is, how well it converts the available dc operating power to signal power, the gain of the dish used by the satellite, the distance to earth, the shape of the transmitted signal, and possibly weather conditions. The FM modulated audio and video signal handled by the transponder is delivered to a small antenna which radiates that signal. If it were not for the dish behind it, that signal would be radiated omnidirectionally into space. The satellite dish concentrates the signal, sending it earthward. At the transponder's antenna, the signal energy is highly concentrated possibly occupying just a few inches. By the time it reaches the earth that energy is spread over the continental U.S. and beyond. Because of this diffusion of the transmitted signal energy, the amount reaching the dish is almost non-existent.

## SPACE LOSS

The transmission of any signal in space results in a loss known as spreading loss or space loss. It isn't confined to downlink signals but is a characteristic of all signals radiated in space whether from earthbound stations or satellites. It affects the uplink signals as well as those that are downlink.

Space loss is a function of distance and is a problem even with terrestrial television stations. Even though these use much higher transmitting powers and much shorter distances, fringe area reception is about

75 miles. Space loss varies as the distance squared, so downlink signals do face a serious transmission problem. Not only is the transmitting power extremely low, but the distance is extremely great, far exceeding that of terrestrial TV stations. Thus, space loss for the downlink signal could be measured in terms of between 100 dB and 200 dB. 100 dB is equivalent to $10^{10}$ or a decrease of the transmitted signal by a factor of 10,000,000,000. However, there is some compensation in the fact that while the operating power of the transponder may be only 5 watts, the radiated signal energy is effectively increased by the gain of the transponder's dish. If this dish has a gain of 50 dB, then the transmitted signal has an effectiveness that is 100,000 times as much. Still, when the signal arrives on earth, it isn't much above the level of the noise floor.

Satellite signals not only travel at the speed of light, but in straight lines, and so do not change their direction unless forced to do so by some object. To be able to receive signals from a particular satellite, a parabolic dish should be pointed straight at it. However, since each satellite may be equipped with 12 or 24 transponders, it is possible to pick up a maximum of 12 or 24 channels from a single satellite without altering the position of the dish, assuming, of course that all the channels are active.

This does not mean the signal from the satellite is reaching earth as a narrow beam, but it does mean the signal radiation covers a large area of the surface of the earth, approximately 40%. A dish located at a distance of hundreds and hundreds of miles from you can pick up the same satellite by following exactly the same procedure—pointing the dish directly at the satellite.

## THE NOISE FLOOR

We live in an extremely noisy universe. Fortunately, our hearing range is so limited we can't hear most of it, but a TVRO can sometimes make it visible.

The energy that the sun sends to earth sets molecules in motion, molecules of air and all other substances. This molecular motion isn't quiet, and the higher the temperature, the greater the noise level. The amount of noise that can be measured is known as the noise floor. Ideally, for a TVRO we would want the received signal level to be as high above the noise floor as possible.

Not only are the earth and its atmosphere sources of noise, but so are all TVRO components, whether passive or active. The flow of electrical currents through various parts of a component is also a noise producer, so TVRO units can be rated on how much noise they generate.

## THE DISH

The dish (Fig. 4-1) is a component that intercepts downlink signals. It is a passive device. It is not an electronics unit, but is the first (and extremely important) part of the TVRO system. It does, on earth, exactly

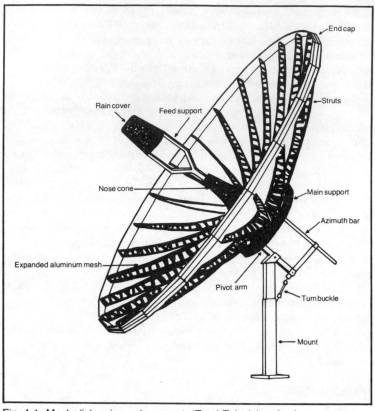

Fig. 4-1. Mesh dish using polar mount. (Total Television, Inc.)

what the dish does for the satellite's transponder—it concentrates the signal.

## SHAPE OF THE DISH

A dish can have any shape you can imagine. It can be round, square, or rectangular. It can be small or large, made of solid metal or wire mesh (Fig. 4-2). Each of these have their advantages and disadvantages. The most commonly used shape is parabolic, although circular dishes are also used.

The outer rim of a dish is circular hence the dish may be called a wagon wheel. The path from any point on that outer rim to a directly opposite point on that rim is usually a parabola.

A parabola can be produced in a number of ways (Fig. 4-3). It can be developed by an equation or it can be generated geometrically. One method is to start with a right cone, a shape similar to that of an ice-cream cone, and to send a plane through it. The points of intersection between the plane and the cone result in a curve that is a parabola.

A plane is just a two-dimensional geometric figure, such as a square or a rectangle. The type of curve that will be produced could be a circle, an ellipse (Fig. 4-4), a hyperbola, or a parabola, depending on the angle of inclination of the plane. Further, any number of parabolas can be produced, all of different sizes, depending on the location of the plane. To obtain a parabola, the plane must be made parallel to the slant edge of the cone as shown in Fig. 4-5.

Fig. 4-2. 11-foot dish using petallized mesh structure. Unit is polar mount type. Designated as the X-11, the dish can be assembled in 2½ hours. It consists of ribbed sections and individual slide-in mesh panels. All hardware is stainless steel and installation can be done with ordinary tools. The tangential drive system eliminates problems of excessive play, limited travel, and the poor leverage common to jack-screw type designs. The heavy gauge steel is gold zinc plated for all-weather protection. (KLM Electronics, Inc.)

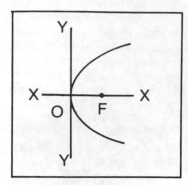

Fig. 4-3. Generalized shape of a parabola. F is the focal point. Lines Y and X are perpendicular to each other.

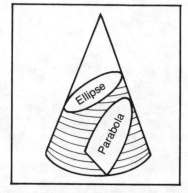

Fig. 4-4. Both the ellipse and parabola are produced by sending a cutting plane through a right cone.

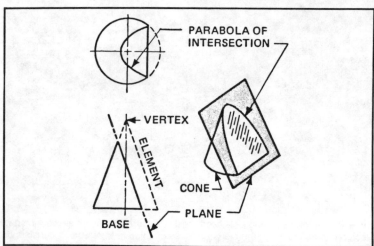

Fig. 4-5. Formation of a parabola. A cutting plane is sent through a right cone parallel to the slant edge. The ends of the parabola rest on the circular base. Thus, a dish may be parabolic, but the ends of the parabolas that form the surface of the dish are circular. Looked at head on, the dish appears circular.

While the shape of a parabolic curve can be obtained geometrically by sending a two-dimensional plane through a right cone, it can also be derived mathematically using the formula: $Y^2 = 4f \times D$. Y is the distance of the curve from the center, f is the focal length, and D is the diameter that is wanted.

It isn't possible for a dish to have any part of its surface *flat* and still be a parabolic dish. A parabolic dish can mean only one thing—it produces a parabolic curve if you move from any point on its perimeter to any opposite point on that perimeter. This means it is parabolic in every direction, moving across the face of the dish, no matter where on the perimeter you start. By definition, mathematically or geometrically, a parabola is a special kind of curve.

The curvature of a dish can be parabolic or spherical, but you cannot determine which it is just by looking at it. A spherical dish is one whose area is cut from the surface of a sphere. That surface section can have a perimeter that is round, square, or rectangular, which is also true of the parabolic dish.

The beginning of a parabolic shape, such as that of a parabolic dish starts as a curved line that is a parabola. When this curved line is rotated about its horizontal axis, it produces a parabolic shape, that is, a device whose surface consists of an extremely large number of adjacent parabolic lines.

A common example of a parabolic surface is that of the reflector of an automobile headlight, as shown in Fig. 4-6A. The light from a source,

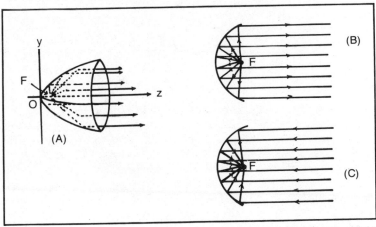

Fig. 4-6. Auto headlight (A) uses reflector having parabolic shape. Lines of light from point source F strike parabolic reflector and are directed outward in straight lines. The reverse action takes place in a parabolic dish. There the satellite signals strike the reflector and are focused to some point, such as F. The parabolic dish can be used for signal transmission. The microwave signal source is positioned at the focal point F (B). For signal reception the microwaves are reflected by the dish and are picked up at the focal point F (C).

marked f, travels in straight lines away from that source. Some of the light strikes the walls of the reflector and then travels outward along straight, parallel horizontal lines. Figure 4-6B shows how the parabolic dish is used for signal transmission. The dish works in a similar, but opposite manner for signal reception. The horizontal lines represent the signal being received from satellite. The signals strike the walls of the dish and are then reflected to a meeting point, or focal point. At the focal point, there is a signal concentration, representing the sum of all the signal voltages bounced off the wall of the dish and reaching the focal point. The signal intensity at point f is greater than that at any other point in or around the surface of the dish.

## THE LENS CONCEPT

The word *lens* is commonly associated with glass made to have a special shape such as those used in binoculars, telescopes, and microscopes. In these instances, the lens works as a light gathering device, concentrating and focusing the light. A lens need not necessarily be made of glass. In a cathode-ray tube such as that used in test instruments or a television picture tube (actually a cathode-ray tube made especially for this purpose), the electron beam produced in the tube is focused electrostatically or electromagnetically, hence the elements that do the focusing are referred to as a lens.

The same concept can be applied to a dish. The purpose of the dish is not only to collect the signal energy, but to focus it, and for this reason the dish is sometimes (not often) referred to as a lens.

## APERTURE

This is the diameter of the dish and is the distance measured, in feet or in meters, from any point on the perimeter of the dish to a diagonally positioned point that is directly opposite, assuming the perimeter is a circle.

Aperture does not take parabolic depth into consideration. For a dish having a circular perimeter, and this is applicable to parabolic dishes as well as spherical, the aperture should be a constant and should be the same no matter what points on the perimeter are measured.

As a general rule, the larger the diameter of the dish, the better. Dishes used in connection with uplink signals are quite substantial and uncommonly have apertures of approximately 30 feet (9.1 meters). Dishes used by cable companies are required to have a minimum aperture of at least 14.76 feet (4.5 meters).

## DISH MATERIALS

Dishes are made of a variety of materials including stainless steel, aluminum, and wire mesh. In some instances, the metallic body of the dish is covered with a coating of fiberglass (Fig. 4-7) but this is simply to protect

Fig. 4-7. Fiberglass dish is transported in sections to the installation site where it is assembled. (Channel Master, Div. Avnet, Inc.)

the metal shell. It is the metal that is the reflecting surface, not the fiberglass.

Dishes can also be coated with special materials to keep their surfaces from oxidizing. Oxidation results in pits which produce a deformation of the metal surface. The coating that is selected is a type that is absorptive rather than reflective so as to avoid any possible concentration of heat at the entrance to the low-noise preamplifier.

The effectiveness of a dish depends on how accurately it is manufactured. Unfortunately, many dishes look alike and there is no way anyone can determine the parabolic curve accuracy of a dish just by looking at it.

The need for absolute parabolic curvature of the dish stems from the fact that the focal point, the entrance to the feed, is small. Deviation from parabolic curvature means some part of the signal is reflected to an area outside the focal point, and to that extent some part of the overall signal is lost. This may not sound like much, but the signal is extremely weak and no part of it can be sacrificed. Further, at this stage, the signal is not only in strong competition with thermal noise but is literally buried in it. At this time the noise level is greater than the signal level.

Dishes can be made of solid metal or wire mesh. Wire mesh has a number of advantages. It is lighter than a comparably sized solid dish and is less subject to wind force. It isn't as prone to the forces of contraction and expansion due to weather. Further, it can be put together in petal form assembly at the viewing site, although some fiberglass dishes are now using this technique.

The wire mesh can be *painted*, an unfortunate word since it gives the impression that such dishes are brush covered. The method of *painting* wire mesh dishes is an involved electrochemical process in which the coating substance, available in different colors, is literally bonded to the mesh, giving it a smooth, uniform, weather-protected surface. Fiberglass can be used as an alternative, but this adds to the weight of the dish. However, fiberglass has much to recommend it and many dishes are made using it. It is important for this coating to be manufactured properly. One of the problems is that the fiberglass must be cured uniformly over its entire surface. If this isn't done it is possible for the setting of the fiberglass to warp, extending that warpage to the reflective material it encloses.

The fact that a dish is made of wire mesh does not mean that signals received from satellites will *leak* through it. The dish is not a sieve and the microwave signals from satellites aren't liquid. There is no analogy here. As far as satellite signals are concerned, they *see* a mesh surface as a solid reflector.

## DISH STRUCTURE

A dish can be a unitized structure, that is, made of a single piece or it can be petallized. A petallized dish is one that is made of sections that resemble petals. The number of petals varies from one manufacturer's dish to another, and both fiberglass and wire mesh types can use this format.

Petallized screen dishes have a number of advantages. Since each section or petal of the dish is a separate unit, packing and shipping the dish is far more practical than shipping a dish that is a single, prefabricated unit.

There are other advantages. The single prefab dish, depending on its diameter, can be quite heavy and difficult to handle. Unless the dish is carefully packed (adding to the already high shipping weight) it is possible for the dish to warp.

Since the petallized screen dish is a *see through* type, its weight is less, making on the spot assembly simple. No particular expertise is required and the dish can be assembled without requiring prior experience. Petallized dishes are available in different colors so these can be selected either to fit in with the environment or to satisfy the personal color tastes of the user. Unobtrusive colors can be selected to ward off possible complaints from neighbors.

## WIND LOADING

Every dish is subject to the force of the wind and must be able to tolerate wind gusts without being moved out of position. Dishes that have a screen structure reduce wind loading. Further, unlike solid structures that can accumulate water, making them unsightly, or that can turn to ice in freezing weather, the screen type does not permit water collection.

## THE DISH AND PROPERTY VALUES

A dish increases property values, just as a swimming pool or any other type of improvement. However, in the event of the sale of a home, removing a solid dish and transporting it may be difficult, if not impossible. Even if transportation should be feasible, it takes as much moving expertise to transport a large, solid dish as it does to move a baby grand piano. Dishes that suffer damage in transportation can either lose some of their effectiveness or become completely worthless.

The petallized screen dish, however, can be disassembled as easily and as quickly as it was put together, and while each petal should be handled carefully, does not require the use of an oversize truck.

## MICROWAVE RADIATION

To some people the appearance of a dish may have a frightening aspect, but this is only because they may look on it as something new and different. Telephone companies make frequent use of dishes, but since these are mounted on towers, they do not seem to pose a threat, even though those towers may have not one but a cluster of dishes. In some instances the proposed installation of dishes has been fought with the argument that a dish enhances the possibility of dangerous microwave radiation. A similar argument was once used in connection with television receivers and in-home microwave ovens. Both of these are now common household items.

A dish does not radiate microwaves. Microwaves are all around us, not only supplied by satellites but because microwaves are used for communications. The charge that a dish is dangerous because of microwave transmissions is complete nonsense.

## CIRCULAR APPEARANCE OF A PARABOLIC DISH

It is difficult to look at a dish and to realize the dish has a parabolic shape, not spherical, particularly when viewing the dish by facing it directly.

Figure 4-8 shows the relationship of the circular edge of the dish with reference to its parabolic surface. As indicated in the illustration, the circle and the parabola are at right angles to each other. Thus, when the dish is made to face a satellite, it is the parabolic surface that does so and this is its most signal receptive position. The circular portion is at right angles to the signal and so is least affected by it.

## DEPTH OF A DISH

The shape of a parabola can be obtained in several ways. It can be secured geometrically, as indicated earlier, by sending a plane, a hypothetical two-dimensional geometric figure through a right cone, or, it can be calculated from a simple equation, $y^2 = 4ax$. All graphs of quadratic equations have the same basic shape which we call parabolic. While the shapes are basic, not all parabolic dishes have the same depth.

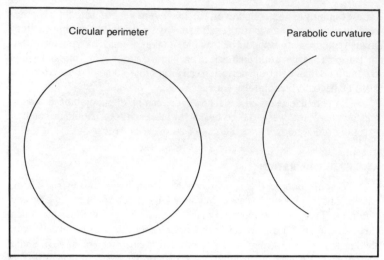

Circular perimeter                    Parabolic curvature

Fig. 4-8. Circular perimeter of a parabolic dish is at right angles to its surface curvature.

## DISH FREQUENCY COVERAGE

Television antennas such as those used for receiving earth-bound broadcasts behave like broadly resonant circuits. This means they can be tuned and this is done by cutting them to a specific length. Since earthbound television antennas must respond to a wide range of frequencies, they are described as broadband types. The length of the TV antenna is cut so as to be resonant in the approximate frequency center of the VHF band. For UHF the frequencies are much higher and so the antenna is made smaller.

A dish used for a TVRO installation is not an antenna and so does not have a resonant frequency as far as satellite signals are concerned. This means that a dish having a width of 10 feet (approximately 3 meters) can be used not only for the C band (3.7 GHz to 4.2 GHz) but also for the K band (11.7 to 12.2 GHz). The K band is used for direct broadcast satellites (see Chapter 8).

Actually, since a bigger diameter means greater signal gain, a larger dish, such as one having a diameter of 10 feet or more will supply more gain than dishes made specifically for broadcast satellite service. These are projected as having diameters of 2 feet to 4 feet. Consequently, the larger dish will help supply a better picture on the much higher frequency K band.

## DISH SIZE VS GAIN

The diameter, or aperture, of a dish is a factor in signal gain. The larger the aperture, the greater the gain. Dishes having apertures of 2 feet to 4 feet do not supply enough gain for existing C-band satellite reception.

The advantage of the smaller dish, practical only for direct broadcast satellite reception, is lighter weight and a better wind load survival. This means such dishes can be mounted on rooftops provided their weight will not be an excessive load on the roof support.

## WIND LOAD SURVIVAL

The ability of a dish to survive the pressure of the wind is measured in a wind tunnel. Dishes made of mesh are better in this respect than those that present a solid surface to the wind. If the dish has become iced, its wind pressure survival is reduced. Wind pressure specs are supplied with the velocity of the wind given in miles per hour (mph) and with some indication as to whether the test was performed with or without icing conditions.

If the area to be occupied by a dish is noted for gusts of high velocity, one possible approach would be to construct a wind break, provided it did not interfere with the *look* ability of the dish. The table shown in Fig. 4-9 supplies the wind speed in miles per hour at 60 degrees Fahrenheit.

## CORROSION

Since the dish is located outdoors, corrosion is a constant threat. While it is possible to make a comparison with television receiving antennas used for terrestrial broadcasting, there is no way of equating the two. Terrestrial broadcasting produces such a strong signal that indoor antennas such as *rabbit ears* are sometimes used. Some manufacturers of television sets even supply an indoor type folded doublet antenna made of two-wire transmission line which can be tacked along a baseboard. In many areas, such an elementary antenna can bring in satisfactory signals.

No such situation exists with dishes. If the dish is unable to collect and reflect an adequate signal level the picture will be poor. Because of its size, a dish cannot be used indoors. Because of the frequencies used for downlink signals, the dish must have a clear line of sight to a selected satellite. While a line of sight to a terrestrial broadcast station is desirable, it isn't imperative.

Dish corrosion can take place in several ways. One is by oxidation. The materials used in making the dish or the frame that supports it, or the hardware (nuts, bolts and screws) are all subject to rusting. Rusting is a chemical process in which the metal forms a compound with the oxygen in the air. Iron, for example, forms iron oxide, better known as rust. Contrary to popular opinion, aluminum can also oxidize. In the rusting process, metals become pitted and lose their structural strength.

The other type of corrosion is due to electrolysis. If two dissimilar metals touch each other, they form the equivalent of a voltage cell with a measurable voltage between them. This situation is particularly aggravated in areas near the seashore where the atmosphere may contain traces of salt.

Electrolysis changes the metal composition of the two metals and in

| Wind Speed in Miles Per Hour at +60° Fahrenheit-(15.6° Celsius) | Pressure lbs./sq. ft. |
|---|---|
| 20 | 1.6 |
| 21 | 1.75 |
| 22 | 1.925 |
| 23 | 2.1 |
| 24 | 2.3 |
| 25 | 2.5 |
| 26 | 2.7 |
| 27 | 2.9 |
| 28 | 3.2 |
| 29 | 3.4 |
| 30 | 3.6 |
| 31 | 3.9 |
| 32 | 4.1 |
| 33 | 4.4 |
| 34 | 4.65 |
| 35 | 4.95 |
| 36 | 5.2 |
| 37 | 5.5 |
| 38 | 5.8 |
| 39 | 6.0 |
| 40 | 6.5 |
| 41 | 6.75 |
| 42 | 7.0 |
| 43 | 7.5 |
| 44 | 7.75 |
| 45 | 8.0 |
| 46 | 8.5 |
| 47 | 8.75 |
| 48 | 9.25 |
| 49 | 9.75 |
| 50 | 10.0 |
| 60 | 14.5 |
| 70 | 20.0 |

Fig. 4-9. Wind speed and resultant air pressure in lbs. per square foot. A 25 mph wind, with the ambient temperature at 60 degrees F on a dish having a surface area of 70 square feet would be 2.5 × 70 or 175 lbs. of pressure applied to that surface.

some instances causes one of the metals to be consumed. Electrolysis can be overcome by not having dissimilar metals in contact with each other or by coating the metals in such a way that physical contact is impossible.

This doesn't mean a dish will fall apart the day it is set up if such conditions exist. Rusting and electrolysis take time, but inevitably they do shorten the useful working life of a dish.

A dish can be made of any metallic reflective material but this material is usually stainless steel, aluminum, or wire screen. These substances may be used alone or they may be covered with fiberglass. The dishes, like the associated hardware such as nuts, bolts, screws and supports, are made of metal, and since they are constantly exposed to the weather, they are also subject to corrosion. To prevent this, the dish may be treated with a special anti-corrosion substance or a coating process or may be covered with a protective material.

## SHROUD

Sometimes a metallic ring, measuring from 6 inches to 18 inches in width is mounted around the perimeter of the dish. This shroud extends from the edge of the dish, forming a right angle with it. Its function is to keep interference from arriving at the sides.

## THE RECEPTION WINDOW

Moving the dish just a small amount in any direction, vertically or horizontally, will result in a loss of signal. The effect is as though the signal was being received through a small rectangular slot, a sort of rectangular opening or window. The lines of the slot define the area through which signal reception is available. Outside the lines no reception from that particular satellite is possible.

The concept of a window or slot (Fig. 4-10) is significant, for it details the reason why a dish is responsive to just one satellite at a time. Satellite

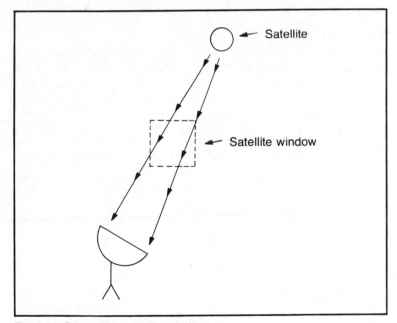

Fig. 4-10. Concept of a satellite window.

signals outside the slot generally do not interfere with the signals from the satellite at which the dish is directly pointing.

While the dish, as explained earlier, is not a tuned circuit, in this sense it can be regarded as the first tuning circuit of the TVRO, a type of mechanical tuning. It works to pick up one satellite, rejecting the others.

## FOCUSING

The purpose of the dish is to focus the signal energy received from a satellite onto a feed. The feed is simply the entrance to a signal conductor known as waveguide. The crossover at the focal point is the maximum signal strength area. Ideally, all the collected signal energy will be reflected to the focal area, but this assumes the dish works as a perfect reflector. Actually, this reflection of the dish is such that while much of the energy does reach the focal area, some of it is scattered. Thus, the signal energy at focus is more like a circular beam, with most of the energy in the center of this beam, and with less signal energy as we move away from the center toward its outer edge. Signal energy is so small that not even a fraction can be permitted to be lost. As indicated in Fig. 4-11 A and B the focal area is surrounded with a scalar feed. This consists of a series of concentric circular rings so designed that any signal energy striking the walls of the rings is reflected toward the focal area.

The focal length of a dish is the distance from the center of the dish to the focal area of the signals. In dish specifications the focal length is related to the diameter of the dish and is written as F/D, that is, the focal length divided by the diameter of the dish. Typical values of F/D are supplied in the table in Fig. 4-12.

Manufacturing specs for dishes do not always indicate focal length but this can be easily calculated by knowing the diameter of the reflector and the F/D ratio. Thus, if a dish has a diameter of 8 feet, multiplying this number by the F/D ratio will supply the focal length. For an 8-foot dish and an F/D ratio of 0.37, the focal length is 8 × 0.37 = 2.96 feet or 35.52 inches.

Usually, the larger the diameter of the dish the greater the focal length, but it is also possible to have two dishes, one with a greater diameter than the other, but with both having the same focal length.

Fig. 4-11A. Feed consists of a short section of waveguide with a series of concentric metal rings near one end. A mounting flange is positioned at the other end for connection to another section of waveguide leading to a low-noise amplifier. (Precision Satellite Systems)

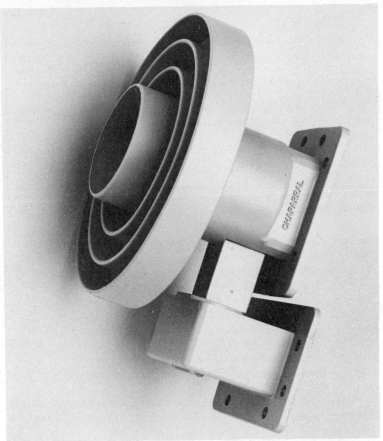

Fig. 4-11B. The advantage of scalar feed is that it reduces signal spillover for minimum interference and noise pickup and gives about 0.5 dB gain improvement over conventional rectangular feeds. (Chaparral Communications, Inc.)

## THE SPHERICAL DISH

Although parabolic dishes are more widely used, spherical dishes do have a number of advantages. As indicated in Fig. 4-13, in a typical parabolic deep dish, the signals from points E and F strike the feedhorn at a high angle and are therefore largely wasted, compared to the signals reflected from nearer the center of the dish, shown as point G in the drawing. This problem is overcome, at least in part, through the use of a scalar feed.

In the spherical dish, as shown in Fig. 4-14A, even the signals from the edge, shown as points H and J, strike the entrance to the feed at a small angle, hence these signals are utilized with good efficiency. The spherical

| APERTURE | FOCAL LENGTH | F/D |
|---|---|---|
| 8 feet (2.5 meters) | 2.91 feet | 0.3638 |
| 10 feet (3 meters) | 4.21 feet | 0.421 |
| 13 feet (4 meters) | 4.9 feet | 0.3769 |
| 16 feet (5 meters) | 4.9 feet | 0.3063 |
| 20 feet (6 meters) | 6.2 feet | 0.31 |

Fig. 4-12. Ratio of focal length to dish diameter. The higher the F/D ratio, the shallower the dish.

dish curvature is obtained by sectioning the surface of a sphere (Fig. 4-14B).

In a parabolic arrangement, the dish must face the selected satellite directly, and so this type of dish must have an arrangement for turning it so as to have a line of sight to other satellites. With the spherical dish, however, the dish does not need to face the satellite directly in order to focus its signal. In fact, the dish can be up to 20 degrees off boresight and still focus a good sharp signal. Since the satellites, on the average, are only 4 degrees apart, this means the spherical dish actually focuses several signals at the same time at different focal points in front of the dish, as shown in Fig. 4-15B.

Unlike the parabolic dish, the spherical dish is fixed permanently in one spot. However, to be able to select more than one satellite, the

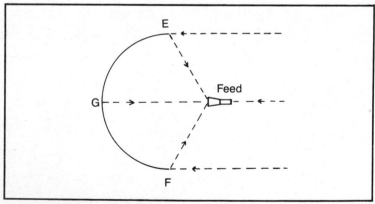

Fig. 4-13. In deep dish parabolic, downlink signals at points E and F, are reflected at high angles to the feed, compared to signals reflected from the center (point G) and are largely ineffective. (McCullough Satellite Equipment, Inc.)

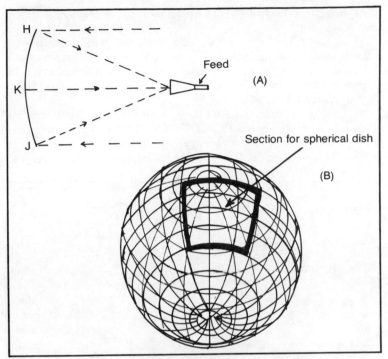

Fig. 4-14A. In a spherical dish, the signals from points H and J strike the feed at a small angle. (McCullough Satellite Equipment, Inc.) Drawing B shows how the spherical dish is obtained by sectioning the surface of a sphere.

feedhorn/LNA must be moved from focal point to focal point. This means a physical adjustment must be made outdoors every time a different satellite is to be tuned in. This may not be difficult to do, particularly if the positioning location of the feed has been previously determined and

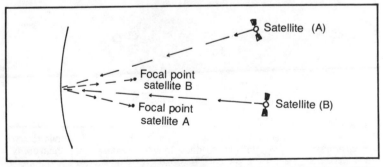

Fig. 4-15A. Spherical dish with pair of focal points for signals from two satellites, received simultaneously. (McCullough Satellite Equipment, Inc.)

85

marked, but it certainly does not have the convenience of indoor control of dish movement.

Because it does not have either a motor operated or manual dish rotating mechanism, the spherical dish can be lower in cost than a comparably sized parabolic dish. Since the spherical dish is a fixed position type, its supporting frame can be made extremely rugged and can be made to have a higher tolerance to wind loading.

The spherical dish is also suitable in locations where the user is unable to pick up more than one or two satellites, or is interested in only one or two, provided the two are adjacent. However, depending on size, some spherical antennas can pick up a number of satellites simultaneously.

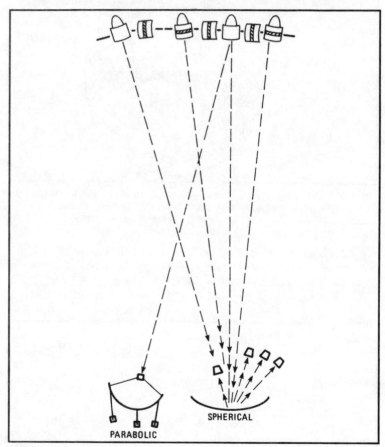

Fig. 4-15B. Comparison of parabolic and spherical dishes. Parabolic focuses signal energy of single satellite onto feed. Gentler curvature of spherical permits reception from several satellites, over 40 degrees of sky arc. A number of different feeds can be used or the dish can be moved to focus on one feed. (Radio-Electronics)

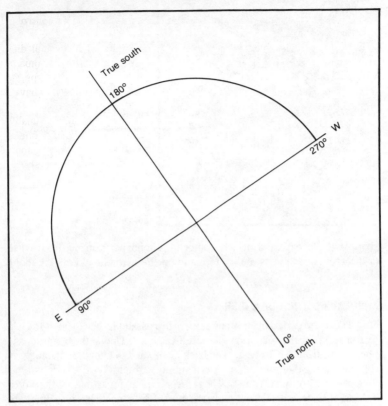

Fig. 4-16A. Azimuth or hour axis. Azimuth is sometimes referred to as a bearing. Azimuth, in degrees, uses true north bearing (not magnetic north) as the reference.

## DISH CURVATURE

One of the most important characteristics of a parabolic dish is its curvature. Thus, all parts of the surface of the dish should be a parabolic section. No area of the dish should be flat. The gain of the antenna is directly dependent on how faithfully the surface area of the dish follows the parabolic curve. There should be no bumps or ridges on the surface, nor should any section be warped or twisted out of shape. Any deformation of the surface area means downgrading of the video signal as seen on the TV screen.

Dish curvature depends directly on a number of factors. The first is the way in which the dish is made. Correct dish curvature must take place at the time of manufacture. Further, the dish must be able to maintain its curvature in all positions, whether horizontal, as it usually is when it is made or almost vertical, as it is when mounted. It must hold its shape during packing and shipping, during unpacking and mounting. It must also maintain its

87

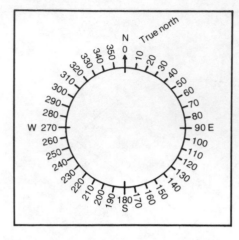

Fig. 4-16B. Angles of azimuth with respect to true north.

shape in the face of wind stress. In short, optimum performance from a dish is directly based on its ability to maintain the symmetry of its parabolic shape.

## ANGULAR POSITION OF A DISH

There are two basic motions used with a parabolic dish to get it to be in line of sight for the reception of a satellite signal. One of these motions is horizontal, from left to right (and back again) and is known as its angle of rotation or angle of azimuth, or more simply, azimuth (or the hour axis) as illustrated in Fig. 4-16A and B. The reference point for azimuth is the North Pole, that is, true north, not magnetic north. True north has an azimuth of zero degrees (or 360 degrees). Using this reference, south is 180 degrees, east is 90 degrees and west is 270 degrees.

The other motion of the dish is up and down and this is known as the angle of elevation (Fig. 4-16C). In this respect, the earth is considered as a

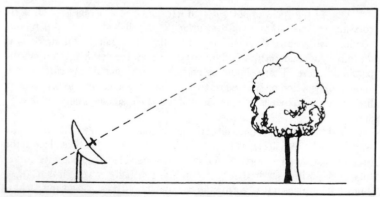

Fig. 4-16C. Angle of elevation. (Total Television, Inc.)

flat surface or zero degrees. An imaginary line drawn vertically at right angles to this reference would be 90 degrees. The angle of elevation for a dish, then, is always some number between 0 and 90 and is specified in degrees.

## MAGNETIC NORTH AND TRUE NORTH

True north is not the same as magnetic north. A compass needle points only to magnetic north. If you could draw a line from the center point of the top part of the earth to the center point at the bottom part, at right angles to the equator, the line would be almost 8,000 miles long. At the top of the earth this line would exit and represent the North Pole and this is the point that is true north.

The earth also has a magnetic north and this is the north to which a compass needle responds. True north is that end of the earth's axis which points toward the North Star and is the point where all meridians converge. The North magnetic pole is in the Canadian Archipelago. The location of the true North Pole is near the center of the Arctic Sea.

The word *bearing* is sometimes used to indicate the azimuth of a dish. Bearing simply means that the dish will be pointing along one of the lines of meridian, and this will be to the left or right of true north. Azimuth figures for dishes are always supplied in terms of true north.

## ELEVATION ADJUSTMENT COVERAGE

Theoretically, elevation adjustment could be from zero degrees, at which point the dish will be parallel to the surface of the earth, to ninety degrees with the dish pointing straight up, forming a right angle with the surface of the earth.

These are extremes and from a practical point of view, are unnecessary. A typical elevation range of a dish could be from 5 degrees to 70 degrees. This does not mean a particular location could have such a wide elevation range, but since elevation ranges vary in different geographic localities, the optimum available range is supplied by dish manufacturers to cover every possible requirement.

## MOVEMENT IN AZIMUTH

The satellites occupy a position in space along an arc parallel to the equator. This arc is an imaginary line, part of a circle, and could be regarded as a spatial line of latitude. It can be marked off in degrees comparable to lines of longitude on earth. The North American satellites occupy a position from 79 degrees west to 135 degrees west longitude. Thus, the dish of a TVRO will face the southwestern part of the sky.

If the dish is made to move so that it scans from Westar 1 at 98 degrees west to Satcom 1 at 135 degrees west it will be moving in a left to right direction, that is, in a direction from east to west (Fig. 4-17). If its direction is reversed, it will then move from right to left or in an easterly direction.

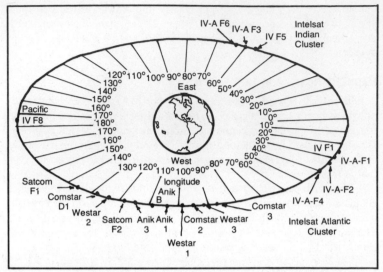

Fig. 4-17. Satellite locations are in terms of degrees, east or west longitude.

Easterly direction does not imply that any of the viewed satellites are located in an eastern orbit. It simply indicates the right to left movement of the dish.

The azimuth angle of movement of the dish is fairly small. This means that in scanning the satellites, dish movement is measured in inches, although the satellite arc is thousands of miles. As indicated in Fig. 4-18 the

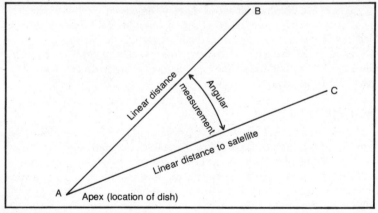

Fig. 4-18. Linear distance depends on the lengths of lines from point A to points B or C. These lines can be lengthened or shortened without affecting the angular measurement. A small movement of the lines using the apex as a pivot point, possibly an inch or so, can result in an increase or decrease of separation of hundreds of miles between points B and C.

angle of an arc is not affected by the length of the lines that form the arc. The angle of an arc can be changed only through the relative movement of one of its lines with respect to the other. We can consider the dish as located at the apex of the arc, the point where the two lines meet.

Angular measurement, such as that used in measurements of azimuth, is not the same as distance. Distance may be in feet, or in miles, or in any other unit of linear measurement, and it becomes greater as we move from the apex of a pair of lines drawn at an angle to each other toward the end of either of these lines. However, the angular measurement remains unchanged. If the angle is 15 degrees for example, close to the apex, it is still 15 degrees at the outer points of those lines.

This concept becomes a little easier to understand by considering a circle. A circle can be regarded as a continuous series of arcs. The total angular measurement of a circle is the sum of the measurements of the arcs that form it and is a maximum of 360 degrees. This applies to circles of all sizes, small, medium, large and tremendous.

## TRACKING

Tracking a satellite is the process of having the dish move in such a way that it follows the orbital path of the satellite. Thus, in tracking the various satellites, the dish will turn in an arc to follow the satellite path. That arc will be circular or elliptical, depending on location. The closer to the equator, the more closely that arc will be toward circular form. As we move north of the equator, that arc becomes ever more elliptical.

Even though the actual azimuth movement of the dish is small, the arc it describes is measured in thousands of miles out in space. A one or two inch left or right traverse by the dish is comparable to a movement of hundreds of miles from one satellite to another. That is why *tuning* a dish, that is, getting it to focus on one satellite, is initially painstaking.

## AZIMUTH ADJUSTMENT

This is sometimes referred to as the tracking range. This indicates that the dish can be positioned so that it faces extremes of horizon, from east to west. On a spec sheet this may be indicated as 180 degrees azimuth adjust. With this amount of azimuth adjustment, every satellite delivering a signal to the North American continent can be *seen*.

However, the fact that a dish has a spec of 180 degrees azimuth adjust, is no guarantee of reception, but is simply one of dish movement capability. The terrain may be such that a clear view can be obtained of only one or two adjacent satellites.

## ELEVATION ADJUSTMENT

The maximum range of this adjustment is from a position with the dish parallel to the surface of the earth or zero degrees to straight up, forming an angle of 90 degrees with the earth.

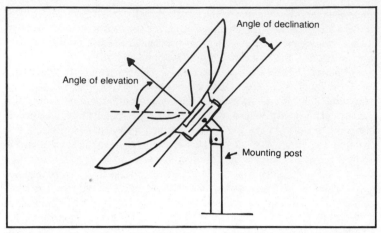

Fig. 4-19. Angle of declination depends on latitude. (Total Television, Inc.)

Such a wide range is unnecessary and so specs will indicate a more practical limit, possibly 5 degrees to 65 degrees. The amount of dish elevation required depends on location. It is less for TVROs positioned in the northern geographical areas, but increases in moving southward.

## DECLINATION/OFFSET ADJUSTMENT

The farther from the equator a polar mount is located (Fig. 4-19) the more it requires a small additional *tilt* adjustment to accurately track the TV satellite orbit. This is *not* an elevation setting for the polar axis but a

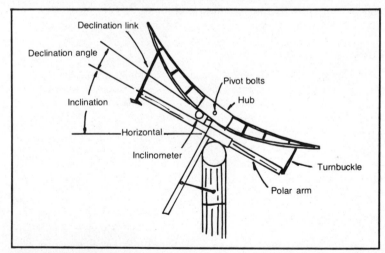

Fig. 4-20. Elevation is sum of declination and inclination angles. (Industrial Scientific, Inc.)

| Lat. | Offset Angle | Lat. | Offset Angle | Lat. | Offset Angle | Lat. | Offset Angle | Lat. | Offset Angle |
|------|--------------|------|--------------|------|--------------|------|--------------|------|--------------|
| 1 | .01 | 16 | 2.8 | 31 | 5.1 | 46 | 7.0 | 61 | 8.3 |
| 2 | .36 | 17 | 3.0 | 32 | 5.3 | 47 | 7.1 | 62 | 8.3 |
| 3 | .54 | 18 | 3.1 | 33 | 5.4 | 48 | 7.2 | 63 | 8.4 |
| 4 | .72 | 19 | 3.3 | 34 | 5.5 | 49 | 7.3 | 64 | 8.4 |
| 5 | .90 | 20 | 3.5 | 35 | 5.7 | 50 | 7.4 | 65 | 8.5 |
| 6 | 1.07 | 21 | 3.6 | 36 | 5.8 | 51 | 7.5 | 66 | 8.5 |
| 7 | 1.25 | 22 | 3.8 | 37 | 6.0 | 52 | 7.6 | 67 | 8.5 |
| 8 | 1.43 | 23 | 4.0 | 38 | 6.1 | 53 | 7.7 | 68 | 8.6 |
| 9 | 1.60 | 24 | 4.1 | 39 | 6.2 | 54 | 7.8 | 69 | 8.6 |
| 10 | 1.78 | 25 | 4.3 | 40 | 6.3 | 55 | 7.8 | 70 | 8.6 |
| 11 | 1.95 | 26 | 4.4 | 41 | 6.5 | 56 | 7.9 | 71 | 8.7 |
| 12 | 2.13 | 27 | 4.6 | 42 | 6.6 | 57 | 8.0 | 72 | 8.7 |
| 13 | 2.3 | 28 | 4.7 | 43 | 6.7 | 58 | 8.0 | 73 | 8.7 |
| 14 | 2.5 | 29 | 4.8 | 44 | 6.8 | 59 | 8.1 | 74 | 8.8 |
| 15 | 2.6 | 30 | 5.0 | 45 | 6.9 | 60 | 8.2 | 75 | 8.8 |

Fig. 4-21. Offset angle (angle of declination) for various latitudes. (Total Television, Inc.)

declination or offset angle between the dish and the polar axis (Fig. 4-20). In Florida or Texas this declination is about 4.5 degrees. For a dish located at the Candian/U.S. border, the declination increases to a little over 7 degrees. Polar mounts that lack this adjustment can accurately track only a small portion of the satellite belt and the quality of reception suffers accordingly.

The tables shown in Fig. 4-21 and Fig. 4-22 list the angle of offset in degrees based on the latitude at your location. Latitude at your location can also be determined by examining the horizontal marker lines on maps and globes, or you can consult an atlas or regional map, or the map shown in Fig. 4-23.

## MOUNTS USED FOR PARABOLIC DISHES

There are three types of mounts used in connection with parabolic dishes: fixed mount, polar mount, and Az/El mount.

### Fixed Mount

The fixed mount is shown in Fig. 4-24. This type of mount is more suited to a spherical dish, but it can also be used for a parabolic dish where reception of only a single selected satellite is wanted or where only a single satellite can be utilized.

The mount used for dishes has two functions. The first and most obvious is that it must support the dish and must hold it in its selected position, under a variety of conditions. Probably one of the worst of these is during the winter, with a heavy wind that sometimes has gusts of up to gale proportions, or with the dish encumbered with snow and ice.

The fixed mount used by cable companies has certain advantages. Once the dish is properly focused, no further adjustments are required. Not

| | Latitude (degrees) | Declination (degrees) |
|---|---|---|
| 25 | | 4.23 |
| 26 | Brownsville | 4.38 |
| 27 | | 4.53 |
| 28 | Tampa | 4.68 |
| 29 | Daytona B. | 4.82 |
| 30 | New Orleans | 4.96 |
| 31 | Hattiesburg | 5.105 |
| 32 | El Paso | 5.24 |
| 33 | San Diego | 5.38 |
| 34 | Atlanta | 5.51 |
| 35 | Albuquerque | 5.65 |
| 36 | Tulsa | 5.77 |
| 37 | Joplin | 5.90 |
| 38 | San Francisco | 6.02 |
| 39 | Washington | 6.15 |
| 40 | Denver | 6.26 |
| 41 | Omaha | 6.38 |
| 42 | Chicago | 6.49 |
| 43 | Caspar | 6.60 |
| 44 | Eugene | 6.71 |
| 45 | Minneapolis | 6.82 |
| 46 | Butte | 6.92 |
| 47 | Tacoma | 7.02 |
| 48 | | 7.12 |
| 49 | Vancouver | 7.21 |
| 50 | | 7.30 |

Fig. 4-22. Declination adjustment for various locations. (KLM Electronics, Inc.)

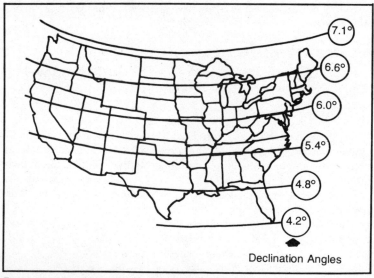

Declination Angles

Fig. 4-23. Declination angles. (Industrial Scientific, Inc.)

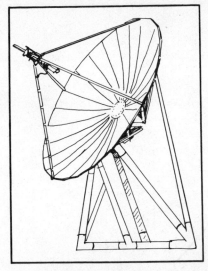

Fig. 4-24. Fixed mount has dish locked into single position. Dish can be spherical or parabolic. Supports can be wood or pipe.

only is the dish support simple, but it can be made to give the dish a higher than normal wind thrust tolerance. Some newspapers that use satellites for news gathering purposes may also use fixed mount supports, a fact that may permit roof mounting. Roof mounting is sometimes necessary when terrain cannot be modified to supply a clear view of a satellite.

### Az/El Mount

The Az/El (azimuth/elevation) mount is so-called since it has two axes of rotation: azimuth and elevation (see Fig. 4-25). These are independent adjustments and while they are often done manually, can be motor operated for remote control.

The problem with making individual azimuth and elevation adjustments is that changing azimuth may also cause a change in elevation, and vice versa. Theoretically, with an Az/El mount it should be possible to make a complete east-west sweep in azimuth without affecting elevation.

The obvious advantage of the Az/El mount over the fixed mount is that it can be used to view more than one satellite. However, the difficulty of making correct azimuth/elevation adjustments limits its usefulness. This type of mount is seldom used in home TVRO setups, but it may be used by cable companies. The usual routine here is to establish a fix on a particular satellite for optimum signal pickup and then not to change the dish's coordinates.

### The Polar Mount

The polar mount, possibly the most popular type used today with home TVRO systems, is illustrated in Fig. 4-26.

The advantage of the polar mount over the Az/El is that just the angle

95

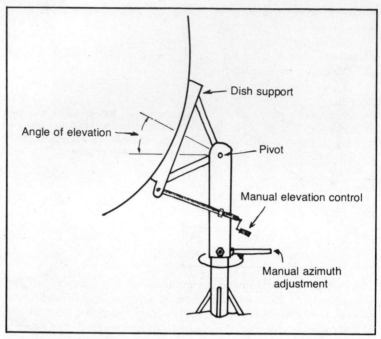

Fig. 4-25. Az/E1 mount.

of azimuth needs to be changed in order to have reception from a number of satellites. The polar mount is so-called since the azimuth motion of the dish is around the polar axis. In this way the dish can be made to scan the orbital arc of all the satellites.

Since the polar mount is supported by its base and since the base and the polar mount are integrated units, they are usually regarded as one.

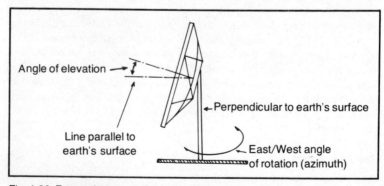

Fig. 4-26. For a polar mount dish two angles must be set correctly. These are the angle of tilt or elevation and the rotation angle or angle of azimuth.

However, the polar mount is that part that supports the dish and enables it to be moved in two directions: elevation and azimuth.

Not all polar mounts are alike. Some will permit tracking just a few satellites, that is, do not permit the dish to *see* all available satellites. Other polar mounts permit a complete sweep.

The dish must be able to track correctly the orbital path of the satellites. A deviation of even a few inches can mean a poor signal, or, quite often, no signal at all. Satellite tracking can be either manual or handled by a motor drive (Fig. 4-27). Usually the motor drive is governed by a control on the satellite receiver.

Fig. 4-27. Rear of petallized dish showing motorized drive assembly with cover removed. The KLM X-11 delivers 40.5 dB gain at 55% efficiency. It has a focal length of 61 inches and a focal-length/diameter ratio of 0.47. Weight is 125 lbs and the wind resistance is up to 100 miles per hour. Windload area is 72 square feet. (KLM Electronics, Inc.)

While motorized control is more desirable, manual tracking could be suitable if the user plans to watch the programs of just a single satellite or possibly two adjacent satellites. The disadvantage of manual control is that the user must make frequent trips to the dish, not only to make the manual adjustments, but possibly some back and forth trips to ensure picking up the best possible signal.

## DISH CONTROL

In some TVRO systems a memory control is used to be able to return precisely to any pre-determined satellite pickup. There are several advantages. With memory control it is possible to return to a selected satellite quickly and accurately. The possibility of undershoot or overshoot is eliminated.

In some installations the dish is fixed in position, and, depending on dish curvature, whether spherical or parabolic, has a field of view toward one or possibly two or three satellites. The Az/El mount can be moved, but for covering the whole arc of satellites, the polar mount is preferable. The polar mount dish can be adjusted manually, but the ideal arrangement is to have a remote dish control equipped with a satellite memory.

## COMPONENT SPECIFICATIONS

Manufacturers supply specifications for the components they supply and this applies to all those that are used in a TVRO system. However, there is no test standardization. This means that any manufacturer may set up any test procedure he wishes and may interpret test results subjectively. Further, manufacturers may also use test instruments made by different companies and these may well have varying orders of accuracy. Even with quality test instruments it is possible to have different results due to day-to-day changes in the results supplied by the same equipment. Finally, there is always the personal element. Operators of test equipment can and do make mistakes and errors are also possible in transcribing test results to printed form.

As a check, there are some evaluations performed by independent testing companies and these are sometimes used to verify a manufacturer's test results. Since tests can be checked, manufacturers are understandably cautious in reporting test results.

Not all specifications are of equal importance and not all spec sheets are alike. In some instances, spec sheets are written on a highly technical, engineering level. Yet, despite all this, spec sheets are valuable, permitting an easy and quick comparison of components. Further, spec sheets may also list the features of units, a bit of information that is useful prior to making a purchase.

Specifications, not only for dishes, but for other TVRO components are often divided into two main categories: electronic and mechanical (Fig. 4-28).

## Electrical

| | |
|---|---|
| Frequency | 3.7-4.2 GHz (C-Band) and 11.7 to 12.2 GHz |
| Polarization | Linear (Fixed, Rotatable 360°) or Dual |
| Gain @ 4.0 GHz ±0.2 dBi= | 40 dB |
| VSWR | 1.25 max. |
| Half Power Beamwidth 1.9° | |
| Input to feed | CPR-229F Waveguide Flange |
| Input to Low Noise Amplifier (optional) | Type "N" female |

## Mechanical

| | |
|---|---|
| Reflector Diameter | 3 meters (10 feet) |
| Mount Type | Polar |
| F/D Ratio | .31 |
| Reflector Surface Tolerance RMS Pointing Error | 040 RMS static 0.06° to 30 MPH winds gusting to 45 MPH and ¼ inch radial ice 0.08° with 45 MPH winds gusting to 60 MPH and ¼ inch radial ice |
| Operating Temperature | -40°C to +55C |
| Survival Wind Loads | 90 MPH winds with ½" radial ice 125 MPH winds-no ice |
| Survival Shock | 1X on Mercalli scale with 30 MPH winds |
| Survival Temperature | -40°C to +70°C |
| Elevation Adjustment Coverage | 10° to 60° |
| Azimuth Adjustment | 0 to 360° |
| Antenna Net Weight | 200 pounds |
| Shipping Cube | 200 cu. ft. |

Fig. 4-28. Typical dish specifications.

## Electronic Specifications

Not all specifications are of equal significance and in some instances the information supplied may seem to be superfluous. One that may fall into this category is frequency range.

## Frequency Range

The frequency range for the downlink band is 3.7 GHz to 4.2 GHz and while this data may be supplied in a spec sheet, it is information of dubious value. All dishes (except those made for the K band) are expected to be able to pick up C band signals. The frequency range is the same for all channels of all satellites.

## Vswr

The purpose of a dish is to collect and deliver the maximum amount of signal energy to the feed. While the surface of the dish doesn't produce signal energy, it can be considered as a signal source as far as the feed is concerned. Ideally, this source delivers all the energy it receives to the feed, with none of it returning. Any signal energy not reaching the feed, and accepted by it, can be reflected to the dish. This unused energy is ultimately dissipated in the form of heat energy, and, in effect, is wasted.

The relationship between the utilized energy and that which is repeatedly reflected is known as the voltage standing wave ratio, abbreviated as vswr. The lower the value, the better. A representative figure would be 1.2 : 1, with 1 : 1 more desirable. The closer the value of vswr to unity, that is, 1 : 1, the greater the amount of signal energy accepted by the feed.

## Side Lobe Performance

It is possible for a dish to pick up competing signals on the same frequency band. The ability of the dish to respond to the wanted signal while rejecting the competing signal is known as its side-lobe performance. If the dish has an aperture of less than 8 feet (2.44 meters) the possibility of side lobe interference is increased.

## Gain

This is one of the more important of the electronic specifications and is directly related to dish aperture. Representative values of gain will range between 35 and 45 dBi. The gain figure should indicate the downlink frequency at which the test was made, since gain tends to be higher at the high frequency end of the C band. However, this information is ordinarily not supplied and if the frequency is mentioned at all it is simply 4 GHz. Ideally, there should be two gain figures: one for the lower frequency end of the C band and one at the higher frequency end.

The larger the aperture of the dish, the greater its gain, obvious since larger dishes have greater surface areas and are capable of reflecting more downlink signal to the feed. Starting with a 6-foot dish the gain will be about 35 dB with an increase of 1 dB of gain for each additional foot in dish diameter. This holds true for dishes having apertures in the 6-foot to 11-foot range. Following that, increases in dish diameter means a gain of only a half dB per diametric foot increase. The chart shown in Fig. 4-29 shows the relationship between the gain of a dish, in dB, compared to dish diameter. The gain of a dish, however, is a function not only of its size, but its shape, and also on how the dish is made.

The gain of a dish also depends on its adherence to its curve, whether that curve is a parabola or spherical. The gain of a dish is also dependent on its accompanying feed. Certain types of feed, as described in the following chapter, can supply a little higher gain, possibly as much as 1 dB.

## NOISE

While gain is an important factor in a satellite system, noise level has equal ranking (Fig. 4-30). Noise can be mechanical or electrical. The sound of a tire against a road is mechanical noise and so is the sound of a hammer striking a nail. For satellite signal reception, this type of noise is inconsequential.

Electrical noise can be produced in many ways. The movement of an electrical current through a conductor or a transistor produces noise, but unlike mechanical noise, it cannot be heard directly. It can be heard through a speaker or it can be seen on the screen of a TV set since electrical noise behaves like a signal. Noise is a signal of irregular frequency whose amplitude or strength varies. Because it behaves so much like an AM signal, it is regarded as a form of amplitude modulation.

Noise, in connection with a TVRO system, has two primary sources. One that is important is thermal energy supplied by the sun. Noise can also be generated by TVRO components. The dish collects two signals. One of these is the desired radiation of signal energy from a particular satellite. The other could be called space noise, due to the fact that the sun is an energy radiator, supplying not only light and heat energy, but energy scattered throughout the frequencies of the satellite C band. The stars are also suns, but because of their tremendous distance, the amount of noise they contribute is small. The earth itself is a noise generator since it is not

| Diameter (Meters) | Gain Of Dish |
|---|---|
| 3.0 | 40.0 dBi |
| 3.7 | 41.8 dBi |
| 4.5 | 43.6 dBi |
| 5.0 | 44.4 dBi |

Fig. 4-29. Gain of dish in decibels, isotropic vs aperture.

| Kelvins | Noise in decibels |
|---------|-------------------|
| 10 | .148 |
| 15 | .220 |
| 20 | .291 |
| 25 | .361 |
| 30 | .429 |
| 35 | .496 |
| 40 | .563 |
| 45 | .628 |
| 50 | .693 |
| 55 | .757 |
| 60 | .819 |
| 65 | .882 |
| 70 | .942 |
| 75 | 1.002 |
| 80 | 1.061 |
| 85 | 1.120 |
| 90 | 1.177 |
| 95 | 1.234 |
| 100 | 1.292 |
| 105 | 1.346 |
| 110 | 1.401 |
| 115 | 1.455 |
| 120 | 1.508 |
| 125 | 1.562 |
| 130 | 1.613 |
| 135 | 1.665 |
| 140 | 1.716 |
| 145 | 1.766 |
| 150 | 1.816 |
| 155 | 1.865 |
| 160 | 1.913 |
| 165 | 1.962 |
| 170 | 2.009 |
| 180 | 2.122 |
| 190 | 2.218 |
| 200 | 2.27 |

Fig. 4-30. Noise temperatures in Kelvins vs noise figures in decibels.

only capable of storing energy but releasing it as well. If this released energy has its spectrum in the C band it will be heard as noise, particularly in areas where the elevation angle of the dish is small.

Noise can be produced by any substance at a temperature greater than absolute zero (zero Kelvins or −273 degrees Celsius). At that temperature molecular motion completely ceases and noise generation stops (Fig. 4-31).

Solid-state devices, such as transistors and diodes, are responsible for a type of noise known as shot effect. This is noise caused by the random arrival of electrons at the terminals of the transistors or diodes. Even

GAsFets (gallium arsenide field effect transistors) produce some electrical noise because of the random arrival of electrons at the drain (the output terminal of the unit). However, modern design of GAsFets is such that the

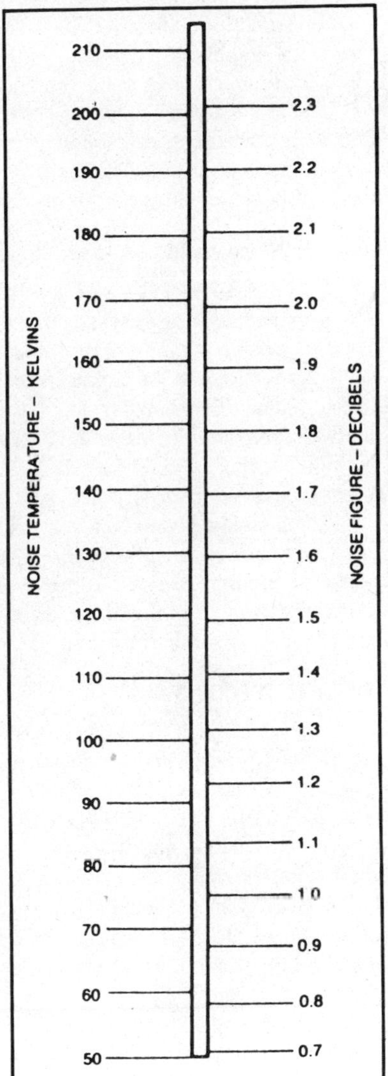

Fig. 4-31. Noise temperature vs noise figure. This data can be supplied in graphic form, as shown above, or in tabular form as shown earlier in Fig. 4-30. (M/A Com Video Satellite, Inc.)

$$dB = 10 \log \left[ \frac{T}{290} + 1 \right]$$

$$T = \left[ \text{antilog } \frac{dB}{10} - 1 \right] 290$$

noise they produce is extremely low compared to other transistor types and for this reason they are widely used in LNAs.

Electrical noise is a signal in its own right. If a noise voltage is delivered to an amplifier, the noise will be strengthened just as a signal would be amplified. This will be straightforward amplification of the noise unless the amplifier is equipped with some means for reducing the signal level of the noise or in some way discriminating against it.

The amount of noise is also dependent on the type of modulation used. One of the advantages of frequency modulation is that it can use circuits that work as noise limiters. These do not eliminate noise but they do cut the peaks of noise signal strength making it less objectionable.

## Signal-to-Noise Ratio

The signal-to-noise ratio, written as S/N, is a comparison between the amount of signal to the noise level, usually measured at the output of a component and expressed in dB. While the S/N gives us an idea of what is available at the output, it does not supply an indication of how well that component has handled the noise in comparison with the signal. For that we would need to know the S/N at the output compared to the S/N at the input. If this ratio, known as the noise figure, and abbreviated as NF, remains constant from output to input, then NF is unity, that is, equal to 1.

Components that have a very wide bandpass, that must be capable of handling a wide range of frequencies, are more subject to noise than narrow bandpass circuitry, for several reasons. A wide bandpass circuit means it encompasses more of the noise range. Further, wide bandpass circuits are often characterized by low gain.

## EFFECT OF A POOR S/N

An evaluation of S/N is more than just an exercise in arithmetic. A low value of S/N means picture degradation. Picture quality may not only be poor but the picture may also be liberally sprinkled with confetti, a word used to describe colored dots or small rectangular streaks. Because these seem to vary in intensity, they are sometimes called sparkle (or sparklies). Sparkle is electrical noise in the chroma (color) part of the received signal.

A video signal, though, consists of two parts: chroma or color, and luminance or brightness (the black and white part of a picture). The monochrome part of the video signal can also exhibit noise and this is in the form of *snow*. Snow appears like black-and-white particles throughout the picture. Like sparkle, snow tends to obscure the image.

Electrical noise not only affects the picture with sparkle and snow but has an effect on the sound as well. The noise is often evident in the form of hiss and is particularly noticeable in the treble range.

It is impossible to eliminate electrical noise completely. The best we can do is to encourage the desired signal and to weaken the noise as much as possible. In a TVRO system this means using components throughout which have an optimum S/N.

The sun is the most prolific producer of space noise, but wind, electrostatic voltages (static) and weather (thunderstorms) are contributors. Noise due to solar radiation is sometimes referred to as cosmic noise.

## FRONT AND REAR OF THE DISH

The front of the dish, that part facing the feedhorn, is a reflector. Since the dish is made of metal, the rear of the dish is also a reflector. Thus, the rear of the dish performs a generally unrecognized function of reducing thermal noise. Thermal noise signals reaching the dish from the rear are reflected in a direction opposite that of signal reflection.

## CARRIER TO NOISE RATIO

This is a comparison of the amount of signal power contained in the downlink signal compared to the amount of noise, with the ratio expressed in dB. Abbreviated as C/N, the higher the value of C and the lower the value of N, the better the picture. 7 dB is about the minimum for C/N.

## NOISE VS DISH ELEVATION

The amount of noise is also a function of dish elevation. In areas where the elevation angle is smaller, the noise temperature is greater. The noise temperature in Kelvins becomes less as the elevation angle increases.

## DISH GAIN AND NOISE

As mentioned earlier, a dish has two important electrical characteristics—its gain and noise. Because of its extremely high operating frequency, the dish isn't bothered with electrical noise signals that can afflict VHF television reception, such as interference from CB and amateur radio signals. The dish is affected by geothermal noise, however.

### Figure of Merit

A dish is a passive device, consequently, it does not amplify the signals it receives from any satellite. However, not all dishes are alike by any means, and some are much better in their signal collecting ability than others. Further, not all locations are alike either, in the sense that some supply more thermal noise than others.

However, it is possible to compare one dish with another and this comparison is known as the figure of merit of the dish. It is directly proportional to the gain of the dish and inversely proportional to the temperature, expressed in Kelvins. The gain of the dish does not refer to amplification, but to its signal collecting ability.

The gain of an ordinary television antenna is simply a comparison with a standard dipole. As far as noise is concerned, a broadcast TV signal is so much greater in strength that any antenna noise that exists is ignored, for the noise signal level is literally swamped by the television signal level.

The same is not true of satellite signals. Such signals have a much different set of working conditions, for what we have here is role reversal when compared to broadcast TV. For satellite signals, thermal noise may be as much as 1,000 times greater than signal level and so noise is a factor that must be considered.

This is expressed as G/T (or figure of merit) in which G is the gain of the dish and T is the noise temperature. However, the dish and its following low-noise amplifier (LNA), although physically separate, work as a unit, and so the noise generated by the LNA is added to the noise of the dish. In terms of a formula:

$$\frac{G}{T} = \frac{gain}{10 \log_{10}} (T1 + T2)$$

What we have here is a comparison, or ratio, of gain to temperature. For best reception G should be as large as possible; T as small as possible. T1 in this equation is the temperature of the dish in Kelvins; T2 is the temperature of the LNA, also in Kelvins.

Dishes having larger apertures, that is, having greater surface areas, have higher values of G since they collect and reflect greater amounts of signal. The value of G is determined not only by dish size but to the extent to which a dish follows a parabolic curve. Any stria or deviations in that surface mean a departure from the curve and affects the figure of merit.

The noise figure, T, depends on location and ambient temperature (you can expect a higher amount of thermal noise on a clear, hot day, in a highly exposed area, with the dish close to the ground). This noise is known as ground noise and sometimes as thermal noise. It is energy developed by molecular motion due to the heat of the sun and is expressed in Kelvins. There is no molecular motion at zero Kelvins but that temperature, if ambient, would not be a life supporting condition.

## DISH BEAMWIDTH

The beamwidth of a parabolic dish, also known as its directivity, is another electrical characteristic. While a dish is not a tuned circuit, it does have the ability, depending on its design and construction, to accept some signals and reject others. This does not mean that the dish can select a particular transponder of one satellite while rejecting all other transponders of that same satellite. Beamwidth is a measure of the ability of a dish to respond to a selected satellite, but not to any others. The beamwidth of a dish is the conical width of the beam projected to it.

It is essential to remember that all satellites use the same downlink frequencies and, since these frequencies are the same, it is possible for a TVRO to process signals from two satellites at the same time. The desired signal may come through strongly, but the unwanted signal may supply picture interference.

In obtaining figures for the calculation of beamwidth, the gain of the dish is first measured with the dish pointing as accurately as possible at a selected satellite. Thus, if a dish has a gain of 40, this is 40 dBi or gain above isotropic, and is made at a measurement of zero degrees azimuth. The dish is then moved off axis, until the gain drops by 3 dB, and the measurement in azimuth is noted. This measurement is done twice, once to the east and once to the west of zero degrees. The total azimuth could be, possibly, 2 degrees. This means that the movement of the dish in azimuth by one degree either way produces a signal attenuation of 3 dB.

Satellites, though, are most often spaced 4 degrees apart, so to get a better indication of directivity, the dish should be turned in azimuth by a total of 4 degrees. The decrease in received signal gain supplies an indication of signal attenuation.

In spec sheets, though, beamwidth is usually measured at 3 dB down, that is, the dish is turned in azimuth until the signal gain decreases by 3 dB. The larger the dish, the narrower the beamwidth and the better the selectivity of the dish.

## NUMBER OF SATELLITES

Theoretically, it is desirable to be able to pick up the signals of as many satellites as possible. For some TVRO owners the thrill of satellite TV is in being able to receive all satellite signals, but to others there may just be the

| | |
|---|---|
| Independent News Network | 21 [W3] |
| Public Broadcasting Systems | 23 [W4] |
| Public Broadcasting Systems | 21 [W4] |
| Public Broadcasting Systems | 17 [W4] |
| Public Broadcasting Systems | 15 [W4] |
| XEW-TV, Mexico City | 06 [W4] |
| Black Entertainment Television | 24 [W5] |
| Daytime | 23 [W5] |
| Alpha Repertory Theater | 23 [W5] |
| SelecTV | 22 [W5] |
| Spotlight (west) | 21 [W5] |
| American Movie Network | 20 [W5] |
| Nashville Network | 17 [W5] |
| Satellite News Channels (feeds) | 16 [W5] |
| Disney Channel (east) | 12 [W5] |
| Satellite News Channels | 11 [W5] |
| Disney Channel (west) | 10 [W5] |
| Satellite News Channels (reg.) | 08 [W5] |
| WOR-TV, New York | 03 [W5] |
| Country Music TV (CMTV) | 18 [D4] |
| Home Box Office (east) | 24 [F3] |
| Cinemax (west) | 23 [F3] |
| Modern Satellite Network | 22 [F3] |

Fig. 4-32. Listing of channel programs. The first two digits are channel numbers. The letters and numbers in brackets are satellite identifications.

choice of possibly one or two satellites that will be important. This is comparable to reception of VHF and UHF channels. Even if all these channels are available, most TV set owners have favorite channels and favorite programs.

As in the case of earthbound TV broadcasting, there are publications available that list programs, channel numbers and the satellites supplying these programs (Fig. 4-32).

# Chapter 5

# Low-Noise Amplifier
# and Downconverter

The low-noise amplifier is so closely associated with the dish that it is often regarded as an integral part of that component. Thus, gain and noise figures are quoted in which the dish and low-noise amplifier are regarded as joined units. Although they are physically separate, electronically they are one.

## THE FEED

In terrestrial TV, the signal is brought from an antenna to the input of the television receiver through the use of transmission line. This connecting link is either two-wire line or coaxial cable.

In a TVRO, the dish focuses the signal at the entrance to a different kind of transmission line known as waveguide. The waveguide is the feed, for it allows the passage of the signal from the focal point, located at its entrance, to the antenna probe, located somewhere near the rear of the waveguide.

## GUIDED VS UNGUIDED WAVES

The utility of a wave lies in the fact that we can move it from one place to another. Such waves can be considered as consisting of two types— guided or unguided. An unguided wave is one that is radiated from an antenna, whether that antenna is a terrestrial TV type or one that is part of a satellite downlink system.

A guided wave is one that is conducted by transmission line (Fig. 5-1) and, for TVROs, that transmission line is coaxial cable and waveguide. Unguided waves, like light, travel through space in straight lines and unlike waves going through coaxial cable and waveguide are unconfined.

Depending on frequency, radio waves can be refracted by the electri-

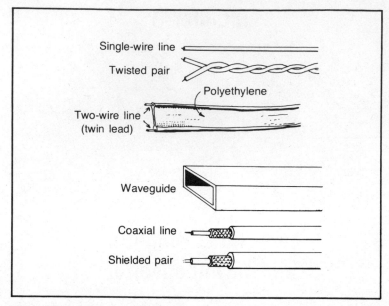

Fig. 5-1. Types of transmission lines.

cal characteristics of the atmosphere, a condition that does not affect radio wave transmissions from satellites. Guided waves, however, follow the contours of the transmission line, and so can be made to change direction, and, in some instances, turn corners.

Unguided waves that strike the reflecting surface of a dish can be concentrated by focusing, but following the focusing point, if allowed to do so, will spread and cover an increasingly greater cross-sectional area because the reflected signals from the surface of the dish do not follow parallel lines.

## WAVEGUIDES

A waveguide, as its name implies, is a device for guiding waves. It is used for the controlled movement of signals from a signal source, such as the reflecting surface of a dish, to some component where it can be processed by a device such as an amplifier.

A waveguide is a hollow metal structure. It can be cylindrical, and look somewhat like a pipe, or it can be rectangularly shaped and have a box like structure as shown in Fig. 5-2. The four walls of the waveguide can be made of solid metal or constructed of small, dovetailed metal sections so arranged that the waveguide is flexible. The purpose of a waveguide is the same as that of the other types of transmission lines, two wire line, and coaxial cable. Waveguide, available in various shapes (Fig. 5-3) is not only essential for the transfer of the signal from the focal point to the antenna probe, but has some advantages as well.

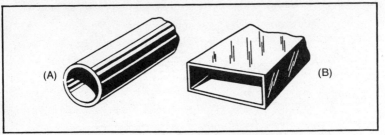

Fig. 5-2. Cylindrical waveguide (A) and rectangular waveguide (B).

## THE TVRO ANTENNA

The signals reflected by the dish are concentrated at the front opening of the waveguide. Instead of diverging following the focal point, the signals, in the form of an electromagnetic field, move through the waveguide. The actual antenna of the TVRO, a small section of metal often U shaped (Fig. 5-4) is centered toward the far end of the waveguide and is immersed in the highly concentrated magnetic field.

In the usual broadcast TV antenna arrangement, the setup for picking up signals is referred to as an antenna, but is actually more than that and should be more correctly referred to as an antenna system. This system consists of the antenna itself, often just a suitable length of rod broadly resonant to the center frequency of the VHF television band, a number of reflectors and directors, and a length of transmission line (either coaxial cable or twin lead). In some instances, especially in fringe areas, there is also a preamplifier inserted between the antenna and the transmission line or the transmission line and the antenna input terminals on the TV receiver.

Satellite signal reception follows a similar arrangement except that the frequencies involved are very much higher. The *TV station* is a satellite some 22,300 miles away and signal strengths are incredibly lower.

Just like VHF reception, an antenna system is required for the pickup of satellite signals. This system consists of the dish, a feed, an antenna probe (this is the actual antenna), and a preamplifier.

The dish is comparable to the reflectors of a VHF antenna; the

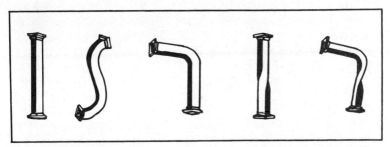

Fig. 5-3. Possible shapes of waveguide sections. Each end of the waveguide terminates in a flange. The input and output impedances are the same.

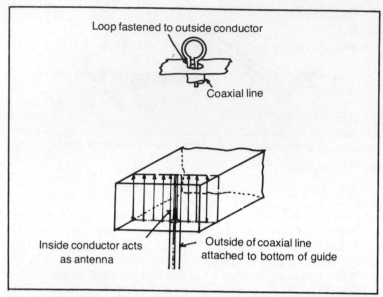

Loop fastened to outside conductor

Coaxial line

Inside conductor acts
as antenna

Outside of coaxial line
attached to bottom of guide

Fig. 5-4. The probe antenna can be a loop, or a straight length of conductor, or can have a rectangular shape. The outside shield braid of the coaxial cable is attached to the waveguide. The antenna probe extends into the inner space of the waveguide.

waveguide is comparable to the twin-lead or coaxial cable. The LNA could be called a signal preamplifier.

Aside from all other factors, there is still one more big difference between a satellite antenna system and a VHF antenna and that is appearance. The dish is large, and the larger the better. Further, even though the dish is just a signal reflector, its shape is far more critical than the parallel rods used as reflectors on a VHF antenna. Not only its shape, but the smoothness of its reflecting surface is important.

## Size of the Probe

Television antennas for VHF have a physical length such that they are resonant in the approximate center of the VHF band. The antenna can be considered as a broadly tuned circuit. The antenna probe positioned in the waveguide immediately prior to the LNA is made physically resonant at the approximate center of the downlink band. Because this frequency is measured in gigahertz, the length of the antenna probe is quite short. The probe isn't just a haphazard length of metal but is carefully measured.

The size of the probe is significant, since its dimensions are such as to favor the signal over the noise. Noise signals whose frequencies are inside the downlink band will be picked up for delivery to the LNA along with the desired signal. The probe will tend to discriminate against noise signals that are above or below the downlink frequencies.

Sometimes the dish is referred to as an antenna. All this means is that the dish and the probe are regarded as being an integral unit. In describing the gain of a dish what is meant is the gain of the antenna probe plus the dish.

Antennas are two terminal devices and so the picked up signal is delivered to transmission line which is also a two-terminal device. In the case of the antenna probe used in waveguide, coaxial cable is used to guide the signal to the input of the following low noise amplifier. The hot lead of the coaxial cable, an unbalanced type, is attached to the antenna probe, while the cold lead or shield lead of the coaxial cable is fastened to the metallic frame of the waveguide.

## TRANSMISSION LINE LOSSES

The use of a waveguide not only offers a convenient way of having an antenna probe for picking up the signal, but is a better method of guiding a wave than open wire line, such as two-wire transmission line or coaxial cable. It has a lower loss than either of these in the microwave frequency range that is used. Waveguide has a lower dielectric loss than coaxial cable, for example. In the case of coaxial cable, the dielectric is the insulating material, a plastic substance, placed between the hot and cold leads. In waveguide, the dielectric is air and air has negligible dielectric loss at microwave frequencies.

## SKIN EFFECT

At the high frequencies used in satellite TV, the current produced by the signal flows in a thin layer near the surface of the conductor, instead of through the entire volume of the conductor. Known as skin effect, it is just as though we had reduced the cross-sectional area of the conductor, thereby increasing its resistance. Consequently, the losses due to resistance are greater than the actual value of resistance as measured with test instruments.

In coaxial cable, most of the loss due to skin effect can be attributed to the inner conductor, the so-called *hot* lead. The reason for this is that its surface area is much less than that of the outer shield braid, which works not only as a shield but as a signal conductor as well.

Suppose we were to remove the inner conductor, keeping only the dielectric material and the outer shield. In so doing we would have a waveguide in tubular form. We could improve its operating efficiency simply by removing the plastic dielectric and using air instead.

Waveguides as transmission lines aren't new. They were used extensively during World War II for radar and still have many uses when microwave frequencies are involved.

## CUTOFF FREQUENCY

Waveguides cannot handle all signal frequencies. At a particular frequency range in the microwave region, the waveguide will have least attenuation, but this increases with frequency. Below the frequency at which the waveguide has optimum signal transmission, it will be unable to function. This non-operative frequency is called the cutoff frequency. Waveguides used for directing the signal from the focal point of the dish to the input of the linear noise amplifier are designed specifically for best transmission of the signal at the downlink frequency.

## SHIELDING

One of the advantages of coaxial cable is the wrap-around shield braid, a feature that protects the signal against unwanted external signals, particularly noise voltages. A hollow pipe or rectangular waveguide has the same advantage, since the metallic shell is not only used to confine and guide the electromagnetic wave of the signal, but keeps external waves out as well. In terms of construction, waveguide is much simpler than coaxial cable. Waveguide does not need connectors of the types used by coaxial cable, and while it isn't as easy to connect and disconnect, can be joined to components by flanges that can be secured by bolts.

## PRECAUTIONS IN USING WAVEGUIDE

If the waveguide should become dented or foreign materials are allowed to get into the guide, signal attenuation can be increased substantially. Since the waveguide is used outdoors, it must be watertight. Waveguides are not fussy about positioning and work well horizontally, vertically or in any in-between arrangement.

The entire waveguide is a so-called *feed*, since it delivers the signal to the antenna probe, but the word *feed* is sometimes used to indicate the entrance to the waveguide which is actually the focal point of the downlink signals.

Like coaxial cable or two-wire transmission line, a waveguide has a certain amount of impedance, depending on its physical construction. Since the signal, in the form of electromagnetic energy, exists at the focal point, this point could be considered to have impedance, known as its open air impedance. For the maximum transfer of signal energy, impedances should match. This means the impedance of the waveguide should be equal to the impedance of free space.

Impedance matching between free space and the input to the waveguide is obtained by flaring the waveguide at its end, as shown in Fig. 5-5. This structure is sometimes called a horn antenna, although it is actually just a specialized form of transmission line.

The flared section of waveguide can be considered as a type of transformer. Its input impedance matches that of the rectangular section of waveguide to which it is attached and of which it forms a part. The flared section has an impedance which approximates that of open space. Since transformers are often used as impedance matching devices, the horn can be regarded as being in this category.

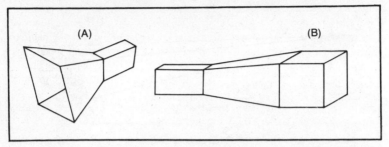

Fig. 5-5. Flared waveguide (A) also known as a horn antenna. The flared section is an impedance matching device. The flared mouth of the waveguide section more closely matches the impedance of open air. Flared section can also be used to impedance match one rectangular waveguide to another (B).

Because of its shape, the flared horn is sometimes referred to as a pyramid. Still another type of horn is the conical, as shown in Fig. 5-6. It is used less often than the pyramid.

## COMPOSITION OF A WAVE

Every radio wave, whether AM or FM, and no matter what its frequency, consists of two components, a magnetic field, usually represented by the letter H and an electric field identified by the letter E. These fields are at right angles to each other and always co-exist, that is, we cannot have one without the other.

A conventional feedhorn may do a relatively poor job of illuminating the dish because the E and H fields coming from the horn *see* different free-space impedances. This is due to fringing or *wrapping around* of the E field. Another way to put it is that the focal point in a horn could be different for the E and H fields. This can be overcome by means of a plate with concentric slots.

## SIGNAL POLARIZATION

The signals transmitted by a transponder can be either vertically or horizontally polarized (Fig. 5-7) A simple way to regard polarization is to consider the electric field of one wave as moving along horizontally, just as though it was a flat sheet of cardboard. The vertically polarized wave can also be imagined as a flat sheet, but forming a right angle with the horizon-

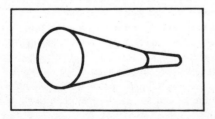

Fig. 5-6. Conical waveguide, also known as a horn antenna.

115

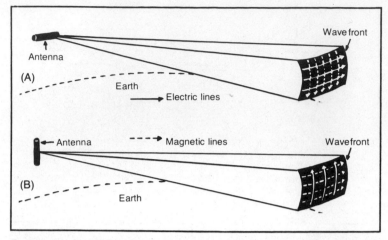

Fig. 5-7. A wave consists of an electric field and a magnetic field at right angles to each other. The polarization of a wave is designated as the direction of the electric field. Electric lines of force (A) and magnetic lines (B). The wave diffuses as it leaves the source. The diffusion of wave energy is gradual as it leaves the antenna and is not as sharp as shown here.

tal. To pick up the horizontally polarized wave one method would be to have the antenna probe in a horizontal position. To pick up a vertically polarized wave the probe would need to be vertical.

In moving away from the source, the signal energy not only diffuses, but moves out in the form of ever-expanding spheres. A small section is a wavefront, but as the wave travels through space the spherical aspect of the wavefront becomes less distinct, and behaves more like a plane at right angles to the direction of signal propagation (Fig. 5-8).

Wave polarization isn't a new technique. In the U.S., terrestrial TV broadcasting uses horizontally polarized waves, hence TV antennas are mounted so that their elements are horizontal. In Great Britain, transmitted television signals are vertically polarized so TV receiving antennas are mounted vertically.

While rotating the antenna probe is a simple way of enabling it to pick up either vertically or horizontally polarized waves, there are actually three ways in which this can be done. The first has already been mentioned and that is by physically rotating the antenna probe. This is an electromechanical method and can be achieved by a small motor combination known as a servomechanism, generally abbreviated as servo.

A typical servomechanism is used to control the rotation of a remote motor shaft. Basically, this arrangement consists of two motors, one in the satellite receiver or other component, the other capable of turning the probe in the waveguide. As an example, if the motor shaft of the control is rotated 90 degrees, the motor shaft connected to the probe would also turn 90 degrees. The amount of rotation, of course, can be less than this.

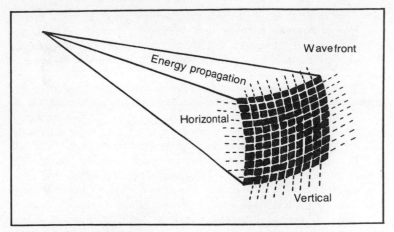

Fig. 5-8. Spherical wavefront.

With this type of polarization control, only the probe antenna turns (Fig. 5 9). The feed and its accompanying LNA remain fixed in position. The servo arrangement is such that the probe can be made to move in extremely small increments, as little as one degree at a time.

The second way of making certain the probe is correct with respect to horizontal or vertical polarization is called the ferrite method and is also known as a dual polarization antenna feed. Such a unit uses the principle of Faraday rotation which results when an axial magnetic field is applied to a waveguide containing ferrite material.

All ferrites are chemical compounds containing oxygen plus a magnetic material. Typically, they are zinc oxide, manganese oxide, nickel oxide,

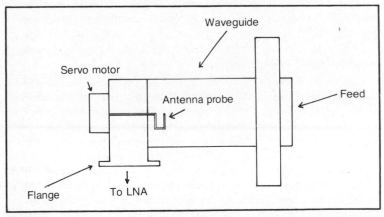

Fig. 5-9. In this arrangement the antenna probe is rotated through an angle of 90° by the servomotor. The selection of vertically or horizontally polarized signals is determined by a control on the panel of the satellite receiver.

and ferric oxide. The ferrite method is used in conjunction with a circular waveguide. With this method, a coil of wire is wound around the ferrite material and a current is sent through the coil. The effect of the resulting magnetic field around the coil is to change the polarization of the received signal.

In application in an antenna feed, the waveguide output is oriented at approximately 45 degrees with respect to the two polarizations to be received. Applying a positive current to the coil then selects one polarization, and a negative current of about the same magnitude selects the opposite polarization. The positive and negative currents can be separately fine-trimmed to compensate for polarization drift resulting from antenna feed alignment or other effects. Switching between polarizations can be accomplished in a few milliseconds. The drive current is quite small and is less than 100 milliamperes for 45 degrees rotation.

The third method is through the use of a switch. In this technique, a pair of antenna probes are used, mounted permanently at right angles to each other. The switch isn't a mechanical type, but uses diodes. These can be made to perform a switching function by turning them on (diode conducts) or off (diode doesn't conduct).

## CROSS POLARIZATION

The vertically and horizontally polarized downlink signals supplied by a transponder in a satellite form right angles with each other. In angular measurement, the separation is 90 degrees. A condition of cross polarization exists when the antenna probe is at the 45 degree point.

## DUAL ORTHOMODE COUPLER

In some instances, a dual LNA is used; one for vertically polarized, the other for horizontally polarized signals. A dual orthomode coupler is a device for delivering both vertically and horizontally polarized signals simultanously.

## SINGLE FEED VS DUAL FEED

A single feed system is one capable of processing satellite signals polarized in one direction only, either horizontal or vertical. The disadvantage is that it has the effect of reducing satellite program delivery by 50% from those satellites that use dual polarity transmissions. Thus, if a satellite has 24 transponders, half of them will be vertically polarized, the other half horizontally polarized. With single feed, only channels that are either vertically polarized or horizontally polarized can be utilized by a feed that is a single polarity type. This disadvantage is eliminated by dual feed, an arrangement in which either vertically or horizontally polarized signals can be processed. Responsiveness to just one type of polarization is sometimes called plane polarization, as opposed to dual. On Satcom and Comstar, both 24 transponder satellites, polarization is odd channel vertical and even numbered channel horizontal. On Westar, another 24 transponder satellite, polarization is vertical for even numbered channels; horizontal for odd numbered channels. A transponder capable of only vertical or horizontal polarization is said to have a half-transponder format. Those capable of both types of polarization, vertical and horizontal, are full-transponder format. Publications listing programs indicate the type of polarization used.

## DUAL AND TRIPLE BEAM FEED SYSTEMS

Single feed and dual feed systems, explained in the preceding paragraph are not the same as dual and triple beam feed systems. Dual and triple beam feed systems permit the simultaneous reception of signals from adjacent satellites. With the dual beam, two satellites with a 4 degree arc can be received. With the triple beam, three satellites within an 8 degree arc can be received.

The dual and triple beam feed systems can work with standard parabolic dish antennas, allowing for retrofit of existing installations in most cases.

## CIRCULAR POLARIZATION

While domestic satellites use either vertical or horizontal polarization (or both), there is still another arrangement known as circular polarization, indicated in Fig. 5-10, a type of polarization utilized by the international Intelsat satellites. As the drawing indicates, the downlink signals proceed from the transponder to the TVRO dish in a sort of corkscrew fashion, with the downlink signals rotating circularly.

There are two types of circular polarization: one in which the signals rotate in a clockwise manner; another in which they turn counterclockwise. By using these two types of polarization for alternate transponders, signals can be downlinked simultaneously on the same frequency. Circular polarization cannot be utilized by antenna probes intended to respond to vertical or horizontal polarization.

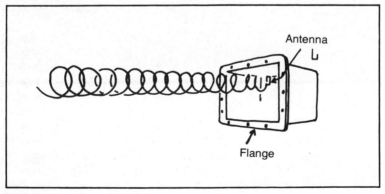

Fig. 5-10. Circular polarization.

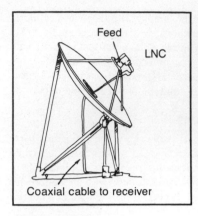

Fig. 5-11. Prime focus feed using a fixed mount dish. The output of the LNC is delivered via coaxial cable to the satellite receiver. The satellite signal is reflected and focused by the dish onto the feed. The feed, a waveguide, delivers the signal to the antenna probe at the input to the LNC (a combination of low-noise amplifier and downconverter). From the LNC, the satellite signal, at a frequency of 70 MHz, is brought to the satellite receiver.

## PRIME FOCUS FEED

A prime focus feed is one in which the feed is positioned so as to be at the focal point of the dish. This is the type of feed most commonly used because it is simple and easy to adjust. In the prime focus feed the LNA is mounted in front of the dish with the feed connected directly to it. Prime focus feed is illustrated in Fig. 5-11.

## CASSEGRAIN FEED

The Cassegrain feed makes use of two reflectors, with the additional reflector located at the focal point. This is the point at which the feed is placed. Cassegrain feed, illustrated in Fig. 5-12A and B is also known as a double reflector, since the dish itself also works as a reflector. Unlike the dish that has a parabolic shape, the second reflector surface has a curve that is a hyperbola. The hyperbolic shape is illustrated in Fig. 5-13 and its

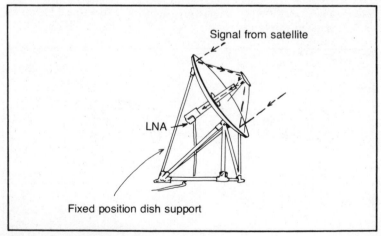

Fig. 5-12A. Cassegrain feed.

Fig. 5-12B. Cassegrain dish using Az/El mount. (M/A Com Video Satellite, Inc.)

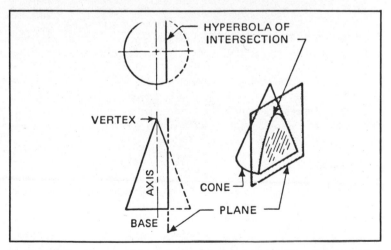

Fig. 5-13. Steps in the production of a hyperbolic curve. A cutting plane is passed through a right cone parallel to the vertical axis of the cone. The points of intersection form a curve called a hyperbola. The plane can be any distance from the axis.

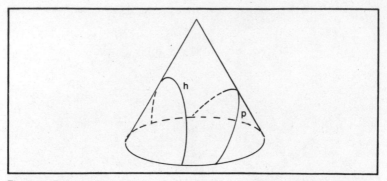

Fig. 5-14. Relationship of the parabola and hyperbola. The parabola is formed by sending a plane through a right cone so that the plane is parallel to the slant edge of the cone. The hyperbola is produced by sending a plane through the cone so that the plane forms a right angle with the base. The parabola is indicated by the letter p; the hyperbola by h. Note that the ends of both curves are on the circular base of the cone, hence reflectors made using either of these curves will look circular when viewed head on.

relationship to a parabola is indicated in Fig. 5-14. The second reflector, because of its shape, is sometimes known as a hyperboloidal subreflector.

In Cassegrain feed, the LNA is mounted at the rear of the dish. The satellite signals reach the subreflector positioned in front of the dish, somewhat forward of its center. The signals are then reflected to the LNA connected to a circular waveguide.

Cassegrain feed is superior to prime focus feed, since it has a higher ratio of dish gain to temperature, but it is more expensive and requires more careful adjustment. However, the prime focus feed is superior to Cassegain feed with respect to side lobe performance. Further, since the subreflector is directly in the path of the signal energy arriving from a satellite, it means less energy directed toward the feed, a loss that is referred to as aperture blockage. The alternative is to increase the size of the dish to compensate. Another method of overcoming aperture blockage is through a special design of both reflectors: the dish and the subreflector. While Cassegrain feed finds application in commercial installations, it isn't used for in-home type TVROs.

## DISH ILLUMINATION

This term, dish illumination, indicates that portion of the dish that is *seen* by the feed. If the feed isn't adjusted properly, it may see more than just the area of the dish. Its *view* of the dish will then extend beyond the perimeter, or it may see only a smaller area that is within the perimeter. Properly adjusted, the feed will see the maximum dish area (Fig. 5-15). Under these circumstances the feed will receive the greatest amount of signal reflected by the dish (Fig. 5-16). The correct location of the feed is at the focal point of the dish.

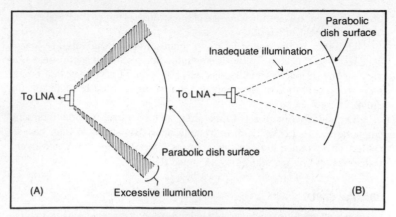

Fig. 5-15. If feed sees beyond the perimeter of the dish it will pick up additional noise but no additional signal, making the signal-to-noise ratio poorer (drawing A). If the feed does not receive signal input from the entire surface of the dish (drawing B), there may not be enough signal for a good picture.

When a feed *sees* beyond the perimeter of the dish, it will pick up additional noise, but it will not pick up additional signal. As a consequence, the signal-to-noise ratio will become poorer. That is why the feed is adjustable and can be moved so as to obtain the greatest signal input from the dish. A condition in which the feed *sees* beyond the perimeter of the dish is sometimes referred to as spillover.

## THE STUB

The antenna probe mounted close to the input of the LNA is some-

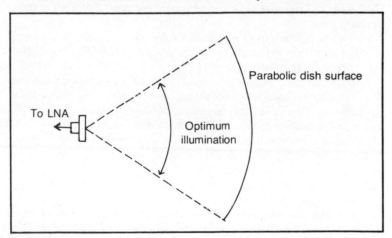

Fig. 5-16. In this setup the feed sees the entire surface area of the dish so signal pickup is maximum.

times referred to as a stub. The size of the stub determines the frequency at which it is most efficient.

The stub is immersed in the magnetic field of the satellite signal conducted to the stub by the waveguide. Since that magnetic field is a varying one, it induces a voltage across the stub. That voltage is a *microscopic* representation of the original signal voltages used to modulate the uplink carrier.

The signal voltage across the stub can be brought into the following component, the LNA, by an extremely short transmission line, coaxial cable. That connecting link must be as short as possible since every transmission line introduces some signal loss.

## COAXIAL CABLE

Like waveguide, coaxial cable belongs to the transmission line family. It is used extensively in TVRO systems for guiding the signal from component to component until it is finally safely delivered to the input of the in-home TV receiver. There are two basic types of coaxial cable: balanced and unbalanced. Of these two, unbalanced coaxial cable is the type used in TVRO setups.

## TRANSMISSION LINES

Whether you use a television set for terrestrial broadcast only or for satellite signal reception as well, there must be some way of conducting the signal picked up by the antenna to the antenna input terminals on the back of the TV set.

For terrestrial television signals, the conductive link between the antenna and the television set can be either two-wire line or coaxial cable. Two-wire line and coaxial cable come under the general heading of transmission lines.

Two-wire line is more economical and easier to install. Coaxial cable has the advantage that its construction supplies a shield against outside electrical interference.

Balanced coaxial cable uses a pair of wires for conducting the signal. The wires are encased in an insulating material which is then surrounded by metallic shield braid. The only function of this outside covering of braid is to act as a shield against external electrical interference.

Unbalanced coaxial cable consists of a single central conductor sometimes known as the *hot* lead. This is simply a descriptive term and has nothing to do with temperature. The *hot* lead is covered with an insulating plastic material, which, in turn, has an outside covering of flexible metal braid, referred to as the *cold* lead.

The braid not only works as a signal conductor in conjunction with the hot lead, but is also a shield against outside electrical interference.

There are a number of different types of unbalanced coaxial cable and Fig. 5-17 illustrates some of these. The electrical characteristics of unbal-

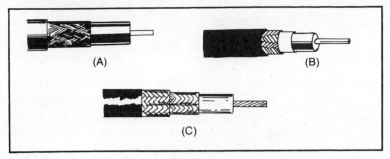

Fig. 5-17. Types of unbalanced coax. The type shown in drawing A has copper braid as the cold lead, surrounded by a black PVC jacket. That shown in drawing B has a foil covering between the braid and the dielectric. Drawing C is coax with a double braid that is silver coated, for minimum noise pickup. (Precision Satellite Systems)

anced coax are dependent on its physical construction. The dimensions of unbalanced coax are also a factor in determining its cost.

For a TVRO system, the two most important electrical characteristics of coaxial cable are its attenuation and characteristic impedance. Attenuation means the amount of signal loss per unit length, such as a foot. Obviously, it is desirable to have as little signal loss as possible, but in turn this must be balanced against the cost of the cable.

Unlike unbalanced coax, siamese cable has several conductors, plus the usual outside shield braid. One wire is used to supply bias voltage to the downconverter, while the other is used for rf (radio frequency) and tuning voltage for the voltage tuned oscillator. Both carry dc voltage. The coaxial cable is the signal carrying conductor (Fig. 5-18).

The coaxial cable linking the downconverter at the dish with the satellite receiver in the home can be run as a supported overhead wire, but it is usually more convenient to dig a trench, about a foot or so deep and to put the cable in it. To protect the cable it is put in polyvinyl tubing with the ends sealed to prevent the entrance of water.

There is also cable available having an extremely tough outer covering and this cable can be buried directly without the use of tubing. The cable can

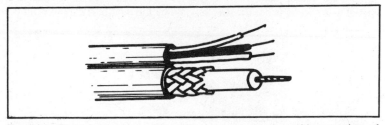

Fig. 5-18. Siamese cable. In this particular example, the cables consists of RG-59 unbalanced coax, plus two-conductor shielded cable with a drain wire (ground lead). Other arrangements are possible. (Precision Satellite Systems)

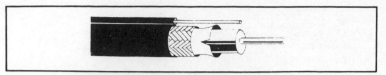

Fig. 5-19. Messenger cable. (Precision Satellite Systems)

be a multi-type (siamese) that is, it may contain coaxial line for the signal, voltage for the motor for the antenna probe, plus other connectors as needed.

Messenger cable consists of unbalanced coax, plus an additional conductor covered with an insulating layer, as shown in Fig. 5-19. This wire can be used to carry a dc operating voltage, working as the plus lead with the shield braid of the coax as the negative return.

Dual coax cable, shown in Fig. 5-20, can be used in installations in which two signals are involved.

## IMPEDANCE

The impedance of a cable is sometimes confused with its resistance, possibly because both are measured in the same unit—the ohm. The resistance of a coaxial cable is a function of its length. The longer the cable the greater its resistance, in ohms. Resistance can be measured with a test instrument such as an ohmmeter.

Impedance includes reactance, the opposition of a conductor to the movement of a varying current through it. Impedance also takes into consideration the frequency of the current. While impedance also includes resistance, generally the amount of resistance is just a fraction of the magnitude of the impedance.

Typical values of coaxial cable impedance are 50 ohms and 75 ohms, values that remain constant regardless of the length of the cable. The impedance of a coaxial cable is no measure of the amount of signal attenuation it may have. A 75-ohm coaxial cable might have more or less signal attenuation than a 50-ohm cable, and this is strictly a function of the length of the cable. The longer the cable the greater the overall attenuation. In effect, the longer the cable the greater the signal loss. How much that signal loss will be is determined not only by the length of the cable but the

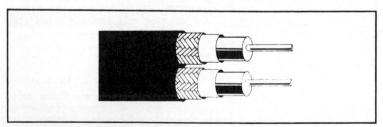

Fig. 5-20. Dual coax. (Precision Satellite Systems)

type of cable and the signal frequency. The higher the frequency, the greater the loss (Fig. 5-21A and B).

## INPUT AND OUTPUT IMPEDANCE

All types of transmission lines, regardless of how they are made, have a certain amount of impedance. Two-wire tranmission line is commonly 300 ohms. This does not mean it is better or worse than coaxial cable rated at 50 ohms or 75 ohms, simply that this electrical characteristic is different.

Not only transmission line, but components such as amplifiers have a certain amount of impedance, but such units have both input and output impedances. They may have a certain amount of impedance at the input, and a completely different value of impedance at the output. As an example, the input impedance of a component could be 100 ohms; the output impedance 200 ohms.

A component, such as a television set for example, can have not only one antenna input impedance but several, as, for example, 75 ohms for coaxial cable and 300 ohms for twin lead transmission line.

## SIGNIFICANCE OF IMPEDANCE

In the absence of fire hoses, a make-do arrangement consists of a bucket brigade, with pails of water being handed from one fire fighter to the next. In a way, a TVRO works along comparable lines. The dish hands the signal over to a waveguide which, in turn, hands the signal over to the stub. In turn, the stub hands the signal over to a short length of coaxial cable which delivers the signal to the LNA.

Because we are working with extremely small amounts of video signal power, it is essential that we lose as little of it as possible. One way of conserving signal power is to match impedances. Thus, the output impedance of one component or cable must match the input impedance of the following component or cable. As an example, if the output impedance of a component is 100 ohms, then that should be the input impedance of the following component. If the output impedance of that component is 50 ohms then it should be connected to a cable having an impedance of 50 ohms or a component having an input impedance, also of 50 ohms.

| DB LOSS | MHz | 57 | 85 | 177 | 213 | 500 | 650 | 800 | 900 |
|---------|-----|-----|-----|-----|-----|-----|-----|-----|-----|
| PER | DRY | 2.1 | 2.5 | 3.6 | 3.8 | 5.9 | 6.8 | 7.7 | 8.0 |
| 100 FEET | WET | 2.1 | 2.5 | 3.6 | 3.8 | 5.9 | 6.8 | 7.7 | 8.0 |

Fig. 5-21A. Typical signal loss figures for RG-59 coaxial cable. These do not apply to all RG-59 coaxial cables made by different manufacturers. Note that the loss, supplied in dB, does not change if the cable is dry or wet. (Channel Master, Div. Avnet, Inc.)

| Type | Jacket O.D. Inch (mm) | Jacket Type | Shield | Dielectric O.D. & Type Inch (mm) | Center Conductor (mm) | Mmfd./ft. (Mmfd./m) | Nom. Imp. Ohms | Nominal Attenuation db | |
|---|---|---|---|---|---|---|---|---|---|
| | | | | | | | | MHz | 100 ft. |
| 59/U | .242 (6.15) | Black Vinyl | Bare copper 95% coverage | .146PE (3.71) | 22 Ga. (.643) copperweld | 21 (68.9) | 73 | 100 200 400 | 3.4 4.9 7.1 |
| 59/U | .242 (6.15) | Black Vinyl | Bare copper 80% coverage | .146PE (3.71) | 22 Ga. (.643) copperweld | 21 (68.0) | 73 | 100 200 400 | 3.4 4.9 7.1 |
| 59/U Stranded Center | .242 | Black Vinyl | Bare copper 95% coverage | .146FPE (3.71) | 22 Ga. 7×30 (.76) bare copper | 17.37 (56.8) | 75 | 100 200 400 | 3.0 4.4 6.5 |
| 59/U cable for 'F' 59 Connectors | .242 (6.15) | Black | Bonded aluminum +60% Alum. braid 100% coverage | .146FPE (3.71) | 22 Ga. (.81) copperweld | 17.3 (56.8) | 75 | 50 100 200 500 900 | 1.8 2.6 3.8 6.2 8.4 |
| 62A/U | .242 (6.15) | | Bare copper 95% coverage | .146SSPE (3.71) | 22 Ga. (.81) | 13.5 (44.3) | 93 | 100 200 400 900 | 3.1 4.4 6.3 11.0 |
| Video Double Braid | .304 (7.72) | Non-contaminating vinyl | Tinned copper double braid 98% coverage | .200PE (5.08) | 20 Ga. (.813) bare copper | 21 (68.9) | 75 | .01 .10 1 4.5 10 100 | .06 .08 .25 .45 .78 2.70 |
| 213/U | .405 (10.29) | Black Vlnyl | Bare copper 97% coverage | .285PE (7.24) | 13 Ga. (2.17) bare copper | 30.8 (101) | 50 | 100 200 400 900 | 2.0 3.0 4.7 7.8 |
| 214/U | .425 (10.80) | Black Vinyl | 2 Silver coated copper 98% coverage | .285PE (7.24) | 13 Ga. (2.26) silver coated copper | 30.8 (101) | 50 | 100 200 400 900 | 2.0 3.0 4.7 7.8 |
| Dual RG/59 Coax Cable | .242 × .505 | Black PVC | Bare copper 95% coverage | .146PE (3.71) | 22 Ga. (.570) copperweld | 20.5 (67.3) | 75 | 100 200 400 | 3.5 5.1 7.5 |

Fig. 5-21B. A few of the more commonly used coaxial cables. (Marshall Electronics, Inc.)

## IMPEDANCE MATCHING AND CONNECTIONS

The fact that the impedance values of two connected components or cables match does not automatically mean the maximum amount of signal will be transferred. Impedance matching is a step in the right direction, but it is just one of several. The connections must be absolutely secure and tight, and when used outdoors must be waterproof as well. Further, all connections must be completely clean with a total absence of corrosion, and must be protected against any such possibility.

## HARDLINE

Ordinarily, coaxial cable is flexible, but there is a type of cable known as hardline that uses a continuous solid metal shield as its cold lead instead of braid. Hardline is used where a rigid connection is wanted between a pair of components.

## THE LOW NOISE AMPLIFIER (LNA)

The signal probe or stub, the actual TVRO antenna, is mounted near the mouth of the LNA, and is the signal source for that component.

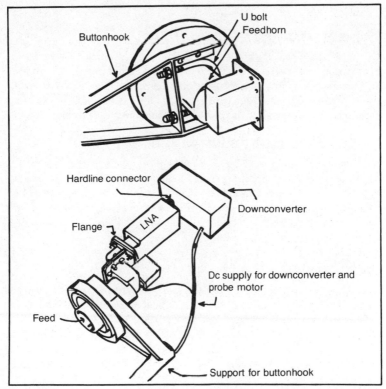

Fig. 5-22. The feed is held in place by a structure known as a buttonhook assembly. The front end of the LNA is flange connected to the waveguide while the output is hardline connected to the input of the downconverter. (Winegard Co.)

The LNA is a broadband non-tunable amplifier and is the first active device encountered by the downlink signal on its way to the in-home television receiver. As an amplifier, its job is to strengthen the extremely weak signal it receives and it does so by multiplying it about 100,000 times. The LNA, however, will not only amplify the signal, but any noise delivered by the antenna probe as well. This is not the only source of noise, for some of it is generated in the LNA itself (Fig. 5-22).

LNAs make use of two types of transistors. One of these at the input side of the LNA is the GaAsFet, commonly referred to as a gasfet. Ga is the chemical symbol for gallium; As for arsenic, while FET is an abbreviation for field effect transistor using gallium arsenide as the doping material. A doping material is a substance that gives a transistor certain desired characteristics. Transistors are almost 100% pure silicon and are then laced with extremely tiny amounts of doping material.

The GaAsFet behaves somewhat like a vacuum tube since it is a voltage, not a current amplifier. It has good linearity and its impedance is stable. Another type that is used is the bipolar transistor.

## UNCOOLED PARAMETRIC AMPLIFIER

This is a more expensive type of LNA. One of its main advantages is that it produces less noise than the usual LNA. Various names are assigned to this amplifier, notably *electronically cooled paramp* and *noncryogenically cooled preamp*. The uncooled parametric amplifier is desirable in situations where the signal level at the input to the amplifier is extremely low.

## CRYOGENICALLY COOLED PARAMETRIC AMPLIFIER

Of the three different types of LNAs this is the most expensive. The amplifier is so called since it is cooled to almost zero Kelvins, corresponding to −459 degrees Fahrenheit. It is a complicated type of amplifier, requiring constant maintenance. Because of its cost and complexity it isn't used in home TVRO's but rather in military installations or in industrial setups.

## THE KELVIN SCALE

Like the Fahrenheit (F) and the Celsius (C) scales, the Kelvin (K) scale is used for the measurement of temperature and is known as the absolute scale. As a comparison, water freezes at 32 degrees F. This corresponds to zero degrees on the Celsius scale; (formerly called the Centigrade scale) and is 273 degrees on the Kelvin scale. A temperature of −273 degrees Celsius, a temperature at which all molecular motion ceases, is equivalent to zero Kelvins.

## LNA RATINGS

LNAs are rated in terms of Kelvins (formerly known as degrees

Kelvin). These could have specifications such as 80, 100, 120 and 150 Kelvins. The most expensive of these are the LNAs that have a temperature specification of 80 Kelvins but most LNAs intended for home use have a rating of 120 Kelvins.

LNA ratings are determined by the amount of electrical noise they generate using zero Kelvins as the reference. If, for example, an LNA has a rating of 100 Kelvins, this means the noise level of that amplifier is 100 Kelvins above a theoretical absolute minimum of zero Kelvins. Thus, the lower the temperature rating of an LNA the lower is noise figure and the better the amplifier.

The LNA, of course, will amplify not only the satellite signal delivered to its input, but any accompanying electrical noise as well. The ratio of signal level to noise level is sometimes written as G/T. In any LNA, it is desirable to have G as large as possible and T as small as possible.

A ratio is a division and so when the division indicated by G/T is completed we have a number known as the figure of merit. The larger this number the better.

There is a relationship between the noise temperature of an LNA and dish aperture. The higher the temperature of the LNA in Kelvins the larger the area of the dish must be to compensate for the increased noise level. While the LNA and the dish are individual components they are integrated in the sense that not only are they physically joined but work closely together.

## LNA NOISE

The noise characteristic of an LNA, that is, the noise generated internally by the LNA, can be expressed in two ways—either in Kelvins or in decibels. The noise difference from one LNA to another as supplied in decibels is quite small as far as numerical values are concerned. Thus, an LNA rated at 120 Kelvins has a noise level of 1.5 dB while another LNA measured at 110 Kelvins has a noise level of 1.4 dB. Noise indicated in Kelvins is always a whole number; in decibels it is often a number having a whole number plus a fractional decimal value. The noise level is more often expressed in Kelvins.

Note, for example, that the difference between an LNA rated at 120 Kelvins is just a tenth of a decibel better than one rated at 110 Kelvins. A tenth of a decibel doesn't sound like much, yet the LNA is probably one of the most critical components in the entire TVRO system. Even a decrease (or an increase) of a tenth of a decibel is significant.

The advantage of having the noise figure in decibels is that the noise produced by other components that follow or precede the LNA can easily be added to that of the LNA to obtain a system noise figure.

There are formulas available for calculating the noise temperature in Kelvins or for determining the noise figure in decibels but these can be obtained much more quickly from the listing shown earlier in Fig. 4-30.

## GAIN OF AN LNA

For most LNAs the gain is 50 dB. This means the signal is 50 dB stronger at the output of the LNA than at the input. 50 dB is a power ratio of 100,000 to 1.

## NOISE FIGURE

Active and passive devices both work against video signals but in different ways. A passive device, such as coaxial cable, a signal splitter, or signal switcher, can have two major faults. The first is loss of signal strength, the other is signal leakage from one channel to another.

An active device can amplify the signal it receives, so the signal strength at its output is greater than at its input. Like passive devices, active components also have some faults. They not only amplify the signal present at the input, but any noise that accompanies that signal. Further, the amplifier also develops internal noise. The internal noise plus the amplified input signal noise are both present at the output. The amount of total noise appearing at the output compared to the amount of noise measured at the input is known as the noise figure of the component, and is measured in decibels.

## GAIN VS THE NOISE FIGURE

The fact that an LNA has a noise figure of 1.5 dB means it will add 1.5 dB of electrical noise to the level present at the input of the component. However, the noise level doesn't affect the gain. Gain is more dependent on frequency. If an LNA has a noise figure of 1.5 dB, for example, its gain could vary by as much as 5 or 6 dB, depending on the frequency.

Gain figures for LNAs are often supplied as a single number, but this is an average, although sometimes a manufacturer will supply the highest gain figure given by his test results. Thus, an LNA could have a gain in the 47 to 53 dB range, but its noise figure could remain at 1.5 dB. The spec sheet for that LNA could indicate a gain of 53 dB but that would be true just for a selected channel.

## GAIN FLATNESS

A spec sheet for an LNA might also indicate the amount of gain flatness. Ideally, an LNA should have the same gain for all channels. The gain flatness figure indicates how much the gain varies from the figure supplied in the specs. In a representative LNA, the gain flatness will not deviate by more than ±0.5 dB.

## THE NOISE FACTOR

The noise factor of an LNA, often represented by the letter f, is the ratio of the signal-to-noise (S/N) at the input to the component compared to the signal-to-noise ratio at the output. Thus:

$$f = \frac{S/N \text{ at input}}{S/N \text{ at output}}$$

Note that f is simply the noise factor and is not the system noise figure. Further, noise factor is not limited to LNAs, and can be used in connection with any other active component. How much a particular unit will affect the system's noise figure depends on its gain. If a component has excellent gain its value of f will override the noise contributed by a following stage. Here we get a condition known as masking, in which the signal gain is substantial enough to swamp or *mask* the noise. The effect is comparable to that of an audio amplifier whose gain is so large that it masks the noise it contributes, and so the noise is not heard, even though it is present. In the case of a video signal, masking would result in the absence of noise on the screen.

## CARRIER-TO-NOISE RATIO

As you can see, because electrical noise is such an important factor in TVROs, it is not only measured, but is compared to the desired signal, not only at the input, but at the output as well. Further, we are interested in its overall effect on the entire system. Not only is the noise compared to the signal, but to the signal's carrier as well, and is known as the carrier-to-noise ratio.

Carrier-to-noise ratio, known as C/N, is the ratio of the signal strength of the modulated carrier from the transponder to the noise level for a selected channel. It is usually measured at the output of the LNA and is given in decibels.

## POWER FOR THE LNA

The LNA is a solid-state amplifier, an active device, and so needs a source of operating power. This power can be supplied in two ways. In commercial installations the LNA may have its own power supply connected to either the 120 volt or 220 volt ac power line. This voltage is stepped down, is rectified and filtered, and is then supplied as low voltage dc to the LNA.

For consumer type TVROs, it is more practical and is safer to have the dc power supply as part of the satellite receiver. The dc operating voltage for the transistors in the LNA is sometimes referred to as bias, a carryover of the same term used in connection with other solid-state amplifiers.

The bias voltage for the LNA ranges from about 15 to 28 volts. The operating current is about 150 milliamperes. Since a milliampere is a thousandth of an ampere, the LNA imposes a rather light load on the power supply in the satellite receiver. Further, since the voltage and current requirements of the LNA are so low, there is no shock hazard.

The dc operating voltage for the LNA remains turned on as long as the satellite receiver is connected to the ac power line in the home, and is continuous whether the power switch of the satellite receiver is in its on or off position.

The reason for constant dc power delivery to the LNA is that this minimizes the possibility of LNA operating range drift. If power to the LNA is turned off and is only turned on when satellite reception is wanted, it is possible that the received picture may be weak, only to improve with the passage of a few minutes or more. This could give the impression of something at fault with the system. The LNA, however, is solid-state and can reasonably be expected to have a long, trouble-free life.

## MULTIPLEXING

While the power supply for the LNA is commonly built into the satellite receiver, it is also possible for the LNA to have a separate power supply known as a power block.

Power can be delivered to the LNA through a separate cable or it can be delivered through the coaxial cable that supplies the signal from the LNA to the receiver. The separate power cable is often color coded to make sure the polarity of the dc voltage is observed. A polarity transposition could damage the LNA. In some instances, polarized connectors are used so as to eliminate the possibility of a polarity transfer.

## LNA CONNECTIONS

There is some standardization in TVRO systems. Thus, the LNA is attached to its accompanying waveguide by a flange (Fig. 5-22). This is a type of mounting plate and it is customary to use one identified as a WR229. The output from the LNA is generally through an N connector with, a waterproof covering and placed at the rear of the LNA housing (Figs. 5-23/5-24).

## LNA AMPLIFICATION

The LNA isn't a tunable amplifier. It is a fixed, broadband type and will amplify all the signals received from the channels supplied by a satellite (Fig. 5-25).

The better the LNA in terms of its amplification, the smaller the possible size of the dish. Ideally, the best arrangement would be to have as large a dish as possible, plus an LNA having the greatest gain and lowest noise level.

## THE DOWNCONVERTER

The signals received by the LNA aren't changed in any way, with the exception of amplification. Except for this factor, the signals at the output of the LNA are the same as those at its input. The frequency of these signals is unchanged and they are all in the gigahertz region. These very high frequencies present a problem since they are so susceptible to signal

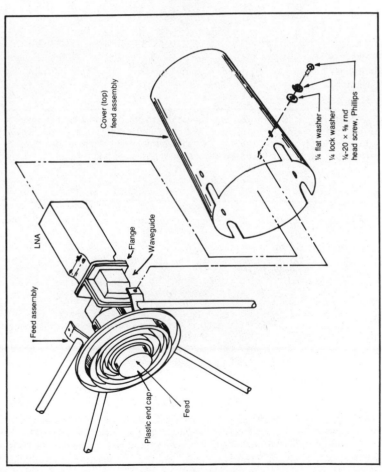

Fig. 5-23. Feed waveguide and LNA assembly with protective cover. (Channel Master, Div. Avnet, Inc.)

Cover (top) feed assembly

¼ flat washer

¼ lock washer

¼-20 × ⅝ rnd head screw, Phillips

LNA

Flange

Waveguide

Feed assembly

Plastic end cap

Feed

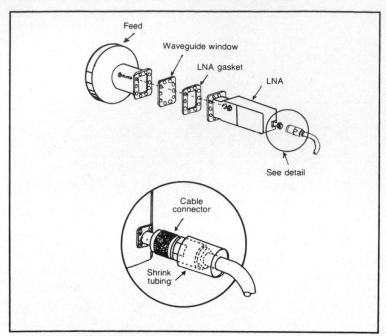

Fig. 5-24. Details of feedhorn, waveguide, and LNA. Shrink tubing supplies a water tight seal at LNA output. Apply even heat to the shrink tubing using a heat gun until the tubing is securely bonded to the cable. If a heat gun is not available, a match or lighter can be used, with caution. (Avantek, Inc.)

Fig. 5-25. Low-noise amplifier is housed in cast aluminum, is equipped with a noise rejection output filter and a lightning suppression circuit. It has reverse bias protection in the event dc voltage of the wrong polarity is supplied from the satellite receiver. (Amplica, Inc.)

136

losses. Thus, if they are routed to another component via ordinary coaxial cable the losses would soon be greater than the gain realized in the LNA.

To overcome this difficulty, the component immediately following the LNA is a downconverter (Fig. 5-26), a unit that can lower the frequency of the signal carrier without affecting the baseband signals, the video and audio modulation.

Downconverters aren't new and have been in use for more than a half century. Every radio and television made today contains a downconverter, except that in radio and television receivers the circuitry isn't referred to as such. All radio and television receivers are superheterodynes, but are based on downconverters.

Essentially, the downconverter used in satellite TVROs and the converters used in radio and television sets works in the same way. Figure 3-2 shown earlier in Chapter 3 supplies the basic concept behind downconversion. The diagram shows two blocks, one labeled local oscillator, the other a mixer. The local oscillator is an electronic generator producing a high frequency continuous wave, very much like the wave used as a carrier. The frequency of this wave can be higher than that of the modulated signal being brought into the mixer, or it can be lower. In either case, both signals being

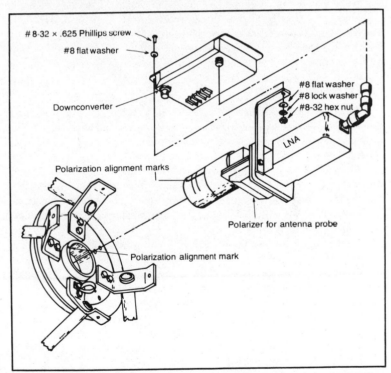

Fig. 5-26. Exploded view of feed, waveguide, polarizer, and downconverter. (Channel Master, Div. Avnet, Inc.)

137

fed into the mixer, that from the local oscillator and that supplied by the LNA, will mix with each other, a process known as beating or heterodyning (from which we obtain the word *superheterodyne*). This mixing or heterodyning process results in a large variety of different signals at the output of the mixer stage. At the output, we have the two original signals, the input signal from the LNA and the signal generated by the local oscillator. We will also have what are known as sum and difference signals, that is, the frequency of the local oscillator signal minus the frequency of the signal supplied by the LNA. We will also have the frequency of the local oscillator plus that supplied by the LNA.

As one example, assume the frequency of the signal from the local oscillator is 4.7 GHz and that obtained from the LNA is 4.0 GHz. The difference signal would be 4.7 GHz minus 4.0 GHz or 0.7 GHz or 70 MHz. The first circuit, and following circuits, after the mixer, are filter circuits designed to reject all frequencies except 70 MHz. This frequency of 70 MHz is known as the intermediate frequency and is abbreviated as i-f. The intermediate frequency circuits not only reject all other frequencies, but are active circuits using transistors, and so, along with frequency selection we get signal amplification.

Note that the input signal has now been changed from one having a frequency of 4.0 GHz to one that is much lower, only 0.7 GHz. The baseband signals consisting of the original video and audio are not affected. So what we have done is to replace an extremely high frequency carrier with one that is lower in frequency, hence more manageable.

## THE SELECTIVITY PROCESS

In an earthbound television system, signal selection is done by two components. A rough sort of signal selection is performed by the television antenna since it works like a very broadly tuned circuit. Signal selection depends almost entirely on tuned circuits inside the television set.

For a TVRO, signal selection is somewhat different. Selectivity starts with the dish, for its azimuth and elevation determines which satellite's transponder signals will be received. Once a particular satellite has been selected, all of the signals of all of its transponders are received and reflected by the dish to its feed.

These signals are conducted by the feed to the antenna probe via a transmission line, a waveguide. This component is not involved in the signal selectivity process. At the end of the waveguide is still another small section of waveguide that forms part of the input to the LNA. This section contains the antenna probe. Here is where the selectivity process continues, for the positioning of the probe will determine which of the 24 possible channels of a satellite will be selected, that is, whether vertically or horizontally polarized signals will be accepted. Thus, the 24 available channels are now reduced to 12.

Now consider the downconversion process. There are 12 signals applying for admission at the input to the mixer. All of these signals will be heterodyned, that is, will enter the mixer and will mix with the local oscillator signal. As a result, a very large number of signals will appear at the output of the mixer, producing an equally large number of intermediate frequencies.

The intermediate frequency circuits are filters and will accept one and only one signal—the one whose frequency produces an intermediate frequency, in this case, of 70 MHz. All the other frequencies will be discarded. Thus, signal selectivity in the TVRO isn't a one-step process, but proceeds step by step until only one signal, the desired signal is selected.

The i-f amplifier section of the downconverter isn't just a single tuned circuit, for just a single circuit couldn't dispose of all the unwanted signals seeking to continue. The i-f consists of a number of stages, one after the other, until every unwanted signal is discarded. This isn't difficult to do since the i-f circuits are fixed tuned, that is, require no operating controls. It is this tremendous selectivity ability that has made the superheterodyne such a superb circuit.

In the example that was supplied, the incoming signal was 4.0 GHz and the frequency of the local oscillator was 4.7 GHz. If another incoming signal is present, and is wanted, then the local oscillator's frequency must be changed. As an example, if the desired incoming signal is 4.2 GHz, then the local oscillator frequency must be changed to 4.9 GHz. The difference frequency would be $4.9 - 4.2 = 0.7$ GHz or 70 MHz. Thus, no matter which input signal is used, the intermediate frequency always remains the same.

## THE VTO

VTO is an abbreviation for voltage tuned oscillator, and, as its name implies, the change in frequency is obtained by varying a voltage with which it is supplied. This voltage is dc and comes from the satellite receiver in the home. To change from one channel to another all that is needed is to depress a pushbutton on the satellite receiver or to rotate a detent type control. In either instance, this will change the amount of dc voltage sent to the VTO and the result will be a change in the operating frequency of the local oscillator, in effect, selecting a different channel.

The voltage delivered to the VTO in the downconverter is typically 8 volts dc. The operating voltage for the LNA is usually 22 volts dc. Both of these voltages are delivered from a power supply in the receiver. The connecting power cable is Belden 8762 or its equivalent.

Because the voltage is dc, the connecting power cable can consist of insulated wire with bare ends. It is much more convenient if the power cable is terminated in spade lugs. This minimizes the possiblility of a wire reaching over and touching an adjacent terminal.

## DOWNCONVERTER DIFFERENCES

The selectivity of the downconverter is important. Since not all downconverters have the same number of intermediate frequency stages, not all of them have the same amount of selectivity. Further, each i-f stage also contributes to the overall gain of the signal. Consequently, a downconverter with a limited number of i-f stages not only has less selectivity but also less gain. These factors help determine whether a picture will be good, simply passable, or even not viewable.

## I-F MODULATION

The type of modulation used for the downlink signals is FM for both video and audio. The heterodyning process in the downconverter performs two jobs: the first is to lower the carrier frequency substantially, and the other is to help select one signal from all those available. The type of modulation is not changed. Thus, the intermediate frequency, like the higher frequency carrier it has replaced, is still FM, for both video and audio.

## THE LNC

The LNA and the downcoverter are often separate components. However, it is possible to combine the downconverter with the LNA, making it into an integration unit. This component is then known as an LNC or low-noise converter (Fig. 5-27). This has an advantage. The distance between the LNA and the downconverter is reduced, consequently signal

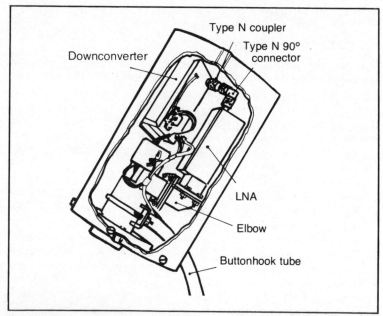

Fig. 5-27. LNC contains LNA and downconverter. Unit is mounted on dish and delivers 70 MHz signals to in-home satellite receiver. (Winegard Satellite Systems)

loss due to a connecting cable is kept to a minimum. Combining the LNA and the downconverter produces a more compact package, lowers manufacturing costs, and simplifies installation.

The downconverter unit, if a separate component, should be placed as close to the LNA as feasible. To connect the LNA to the downconverter RG-142 coaxial cable is used. This is an important connection, since any damaged connectors, crushed or kinked coaxial cable, or a loose connection, can reduce the quality of the signal.

Since the LNA is mounted on the dish assembly, the LNC is mounted there also. Even when the LNA and the downconverter are separate units they are both mounted on or near the dish and so remain near each other.

The output frequency of the downconverter is fairly low, and in the example that was supplied, was only 70 MHz. This frequency can easily be carried by coaxial cable for distances up to 125 feet without serious losses. This means that the signal can be moved from the dish-mounted downconverter or LNC, via coaxial cable, to the satellite receiver, positioned in the home.

## LINE AMPLIFIER

In some installations, a connecting coaxial cable between the downconverter and the in-home satellite receiver may be so long as to introduce intolerable losses. In that case a line amplifier can be inserted in the line to compensate.

## WEATHERPROOFING

If the downconverter is a separate unit, it must be weatherproofed, just like the LNA. Weatherproofing is also required for an LNC.

## THE ISOLATOR

An isolator is a device that permits the passage of signals in one direction while attenuating them in the other. Isolators are used in the connecting line between the output of the LNA and the downconverter.

## BLOCK DOWNCONVERSION

The downconverter that has been described helps supply the signals of a single channel to the satellite receiver in the home. After being processed in the satellite receiver, these signals can be used for one television set, or several. The problem with this arrangement is that all of the television receivers would be tuned to the same channel and the same program would appear on all the screens.

In a number of instances, whether in the home, or in a commercial arrangement such as a motel, or a high rise apartment house, an alternative setup is needed so as to permit individual selection of programs.

The technique used is called block downconversion. With this arrangement, all of the programs supplied by a single satellite are converted, instead of just one. This means that the entire downlink band, ranging from

141

3.7 to 4.2 GHz is converted to an intermediate frequency instead of just one of the frequencies within that band. The frequency range of the downlink band is 500 MHz, that is, 4.2 GHz − 3.7 GHz = 0.5 GHz = 500 MHz. This means that the heterodyning process must produce an intermediate frequency signal that is 500 MHz wide.

The intermediate frequency can be any frequency selected by the manufacturer of the block downconverter. Thus, it could be 500 MHz to 1,000 MHz, or 100 MHz to 600 MHz, and so on. The intermediate frequency amplifier for a block downconverter would need to be a very wide band type.

The use of a block downconverter does not mean 24 channels will be available for single channel selection by a group of viewers using different TV receivers. Only 12 of those channels, either vertically or horizontally polarized, will be picked up by the antenna probe to be amplified by the LNA prior to delivery to the block downconverter.

There is still another factor to consider in connection with this component. The signal from the downconverter is delivered to one or more satellite receivers. These receivers must be designed to accept the wide band signal delivered by the block downconverter.

Finally, the dish, if a parabolic type, will pick up just one satellite at a time. This means that its azimuth and elevation must be under the control of a single operator.

## FEED PRECAUTIONS

The feedhorn and LNA opening must remain clear. Many types of insects (wasps, spiders, etc,) look on the feedhorn and LNA waveguide as the ultimate in modern housing. Remember that any obstruction to the microwave energy will degrade the picture, and if this is noticed, check the opening. In removing any obstructions be careful not to touch the antenna probe. Wasps are particularly notorious, and not only build a rather solid nest, but it is firmly attached to the walls that support it.

The antenna probe inside the opening of the LNA is designed to pick up microwave energy and is positioned precisely to pick up this energy most efficiently. If this probe looks as if it has been bent or pushed over do not attempt to readjust it. The positioning of the probe and its shape are a form of tuning, set at the factory to optimize the performance of the LNA. Depending on what has happened to the probe, you may either receive degraded pictures or no pictures at all. In that event contact your dealer or the manufacturer for advice on how to proceed. Do not disconnect the LNA and ship it back to the manufacturer without prior communications. Some manufacturers will not accept the return of material without authorization.

# Chapter 6

# The Satellite Receiver

After the satellite signal has been changed to a lower frequency by the downconverter, the signal, now with a carrier frequency of 70 MHz, is delivered via coaxial cable to a satellite receiver. Unlike the downconverter and the LNA, this receiver is conveniently positioned near the television set.

The main function of the receiver (Fig. 6-1) is to dispose of the carrier (or carriers) and so, at the output we will have our recovered baseband signals, the video and audio signals originally used to frequency modulate the radio-frequency carrier of the uplink transmission. The process of separating the baseband signals from the carrier is called demodulation, with various types of circuits used for this purpose.

In its travels from the input of the satellite receiver to the output, the signals are passed through amplifiers, so the video and sound signals are constantly being strengthened.

## SELECTION OF AN INTERMEDIATE FREQUENCY

In the example of downconversion supplied in the preceding chapter, the intermediate frequency was indicated as 70 MHz. It is the one commonly used, but the i-f is the choice of the manufacturer, and any other i-f could be used.

## THE OUTDOOR SATELLITE RECEIVER

While the receiver is most commonly placed in the home, in some installations it is mounted on or near the dish. In such cases the operating controls can be remotely adjusted from the home, but this adds to the

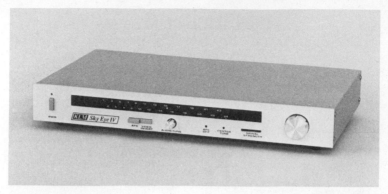

Fig. 6-1. Satellite TV receiver, Sky Eye IV, features slide-rule tuning, signal strength LED bar, center tune LD, afc, video polarity control, fully tunable audio from 5.5 MHz to 7.5 MHz, and remote downconverter. Equipped with SAW filter. (KLM Electronics, Inc.)

complexity and cost. The advantage is that the coaxial cable from the downconverter to the receiver is very short.

## THE FREQUENCY AGILE RECEIVER

For the most part, satellite receivers can respond to all the transponders of a satellite. Such receivers are sometimes called frequency agile (Fig. 6-2) since they can tune from channel to channel. On the other hand, there are some receivers that are dedicated, that is, can pick up one channel

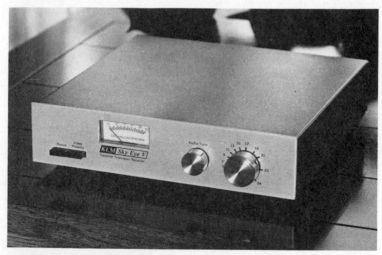

Fig. 6-2. Satellite receiver, (Sky Eye V) has tuning meter, 24 channel rotary control selector, audio tuning, and video polarity pushbutton for inverted video signal scrambling. (KLM Electronics, Inc.)

only. Such a receiver might be used by a newspaper interested in only one satellite or possibly a cable company that works with just one satellite.

## CHANNEL SELECTION

Like any other receiver, one of its functions is to permit channel selection. Since there are 24 transponders per satellite and since each of these can supply a channel, the receiver must have a control system for channel selection (Fig. 6-3). Each is selected by sending a dc voltage to the voltage-tuned oscillator (VTO) in the outdoor downconverter. On the satellite receiver, this is done by manipulating a control. Known as continuous tuning, it is so called since tuning is accomplished by varying the tuning voltage continously.

There are various mechanical methods used for tuning. One of these is known as detent or click stop tuning. The rotary position of the control being used has predetermined stop points that can be heard as a faint click.

Another type of tuning is straightforward variable. In this type of tuning, either the tuning control or the area surrounding it may have marks indicating a particular channel number. To avoid crowding of numbers, a receiver may indicate only the odd-numbered channels. This does not mean even numbered channels cannot be received. These are available for selection between all the numbers indicated.

In some instances, the channel selection control is associated with a digital readout, supplying a quick and highly readable number corresponding to the position of the variable control. Some receivers use an illuminated slide-rule type dial, marked with 24 channel numbers instead of frequency, as in the case of AM and FM sets.

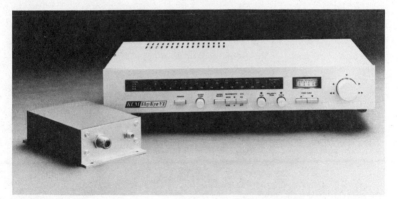

Fig. 6-3. The Sky Eye VI satellite receiver offers VHF-cable/satellite switching that automatically isolates local VHF-cable signals when the unit is in use, preventing co-channel interference. Reception of weak signals is improved by narrow/wide bandwidth control that improves S/N. This receiver has single-knob, fully variable polarizer control, two-speed scan, slide rule tuning and a dial marker that brightens when center-tuned. Receiver is equipped with fully tunable audio, 5.5 to 7.5 MHz. Optional remote control available. (KLM Electronics, Inc.)

145

## SPEED SCAN

Some satellite receivers have single speed scan in which all channels are scanned at a constant rate. The disadvantage of constant scan is that there is so little time to override the scan and make a viewing decision. To overcome this problem, a few satellite receivers are equipped with two speed scan. In the two speed scanning system, scanning slows down when crossing active channels, giving viewers a longer, clearer look. For any channels that are inactive, scanning proceeds at a steady, but not a slow rate.

## CHANNEL CALIBRATION

On the rear apron of the receiver you may find a pair of controls marked *channel calibration*. These will be marked *low* and *hi*. Low refers to low channel numbers; hi for high channel numbers. To calibrate the channel selector on the front panel adjust the *low* control for Channel 3 or 4. Keep working back and forth between these two controls until the 1-24 channel selector on the front panel reads correctly.

As a final check, select Channel 1 and then Channel 24 and see how correct these are. After channel calibration is completed, the channel selector control on the front panel of the receiver should be able to bring in each numbered channel when set to its number.

The channel calibration control is a variable resistance known as a potentiometer, often referred to as a *pot*. Once channel calibration has been adjusted satisfactorily, no further trimming of the pots should be needed.

## CHANNEL TUNING AIDS

You cannot expect the same kind of signal reception from all the satellites any more than you can expect the same kind of reception from all earthbound television stations. Some satellites will produce strong signals; others less so. To get some idea of the comparative strengths of these signals, some TVRO receivers are equipped with meters (Fig. 6-4). This is a feature commonly found on hi/fi receivers but not on television sets.

Various types of signal strength indicators are used. One of these is a signal strength meter and the purpose is to have maximum deflection of the

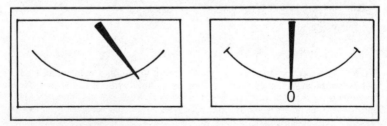

Fig. 6-4. Signal strength meter (left) and zero-center tuning meter (right). These are ballistic-type meters and on some satellite receivers are backlighted.

146

meter pointer. There are no satellite receiver adjustments you can make that will increase pointer deflection, but you can use it to peak the signal by altering elevation and azimuth of the dish or by repositioning the feed. While the picture itself could be used, the meter pointer is much more responsive. Also, with the signal strength meter, you can get some idea of the comparative signal strength of each of the satellites. Still, the signal strength meter is a ballistic device, so even with the use of this meter it may not always be possible to squeeze out the very last bit of signal strength.

Instead of meters, the receiver may be equipped with light-emitting diodes (LEDs) used as indicators. One of these could be used for signal center tuning so that when a satellite is tuned in correctly, the LED will light. A group of such indicators can be arranged in the form of a horizontal bar graph to supply a visual record of signal strength. Comprising five LEDs, maximum signal strength is indicated when all five are illuminated, supplying a horizontal line of light. In some receivers there may be a combination of a ballistic meter and LEDs for center tuning and signal strength.

## SIGNAL SCRAMBLING

Some satellites transmit encoded programs, a technical word to indicate that the program has been scrambled. This is done to keep non-subscribers from viewing the program. The downlink signal, picked up by one or more cable stations, will then be decoded and routed to those who pay for this cable TV service. Alternatively, the cable company may not decode the signal, but will supply decoders to subscribers who are willing to pay for the program.

There are many ways of encoding or scrambling a signal, but one that is commonly used is signal inversion. The present method of video signal modulation is known as negative modulation. This simply means that the stronger parts of the signal produce the blacker parts of the picture. Negative modulation has certain advantages. Thus, the synchronizing and blanking pulses represent the highest modulating signal voltages.

Inverting the composite video signals means it becomes positive. Thus, the blanking and synchronizing signals produce the brightest light on the TV screen. Further, because their polarity has been changed, the picture will not remain in *sync*, that is, it will roll vertically and horizontally. As a result, such a picture isn't viewable.

Some satellite TV receivers have a video inversion feature. This inverts a video signal that has had its polarity transposed, thus producing a viewable signal. However, there are many methods of encoding a television signal, and if inversion is not the process being used, then the inversion feature of the satellite receiver is of no help (Fig. 6-5).

## SINGLE AND DUAL CONVERSION

Figure 6-6 shows a block diagram of a typical AM super-het-

| Satellite | Polarity |
|-----------|----------|
| Satcom II R | Normal |
| Galaxy II | Normal |
| | |
| Satcom IV | Normal |
| Comstar DIII | Normal |
| Westar III | Reverse |
| Comstar D1/D2 | Normal |
| Telstar 301 | Normal |
| Westar IV | Reverse |
| | |
| Anik-D | Reverse |
| Anik-B1 | Reverse |
| Anik-A2/A3 | Reverse |
| | |
| Satcom II | Normal |
| Westar V | Reverse |
| Comstar D4 | Normal |
| Satcom III R | Normal |
| Satcom I | Normal |
| Galaxy I | Normal |
| Satcom I R | Normal |
| Satcom V | Normal |

Fig. 6-5. Satellite signal polarity. This requires adjustment of the polarity control on the receiver.

erodyne receiver. The vertical dashed lines indicate the various sections into which this receiver can be divided. The first section is an rf (radio frequency) amplifier, a high-frequency amplifier that strengthens the received radio signal. The next section is the mixer/local oscillator, an arrangement that downconverts the incoming signal to a new frequency known as the i-f or intermediate frequency. The i-f signal is then brought into a demodulator, a circuit that disposes of the i-f carrier and retrieves the original baseband signal, in this case an audio signal. The audio signal is then delivered to a transducer, an energy changing device such as a speaker or headphones.

The arrangement of Fig. 6-2 is an integrated component since all the sections are mounted on the same frame, generally a printed circuit (PC) board, with the entire structure placed in a single housing or cabinet. However, each of the sections could be a separate component, with each of the components housed individually but connected by cables or other wiring arrangement.

We have a comparable setup with a TVRO. Instead of an rf amplifier, we have a pre-amplifier called an LNA. This is followed by a local-oscillator mixer referred to as a downconverter. The satellite receiver contains intermediate frequency stages with the signals ultimately supplied to a demodulator. However, the TVRO setup does contain one circuit which does not appear in the AM superheterodyne and that is a remodulator.

There are two types of downconverters as described earlier. One of these downconverts just a single channel at a time, producing an i-f, usually

148

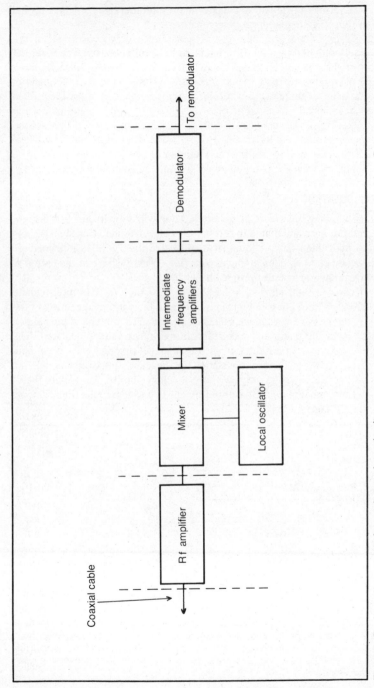

Fig. 6-6. Basic arrangement of superheterodyne circuit.

70 MHz. This i-f is fed into i-f stages in the satellite receiver (Fig. 6-7) and from those i-f stages into a demodulator.

As explained previously, it is possible to downconvert all of the channels supplied by a satellite and to do so simultaneously. The resulting intermediate frequency is not a single frequency, such as 70 MHz, but a band of frequencies that is 500 MHz wide. However, when this band of frequencies is supplied to a satellite receiver, another downconversion step is required, so as to be able to select just one channel out of the possible dozen that are presented. As indicated in Fig. 6-8 this second downconversion results in an i-f of 70 MHz. Known as double or dual downconversion, it is useful when a number of satellite receivers are to be used but with each capable of selecting its own channel for viewing.

## DEMODULATION

Following the i-f circuitry there is a demodulator. Its function is quite straightforward and that is to recover the original modulating signals, both audio and video. In so doing, it disposes of the 70 MHz intermediate frequency that has been acting as a carrier. What we have at the output of the demodulator are the original baseband signals.

It may seem odd not to be able to take the 70 MHz intermediate frequency and to put that signal directly into the antenna terminals of the TV receiver. One reason we cannot do so is due to the type of modulation used. Both the audio and video signals were frequency modulated. The television receiver circuitry is such that it will respond only to an input signal that uses frequency modulation for the audio; amplitude modulation for the video. Consequently, we have no choice but to demodulate the i-f signal and remodulate the baseband signals to conform to the requirements of the TV receiver.

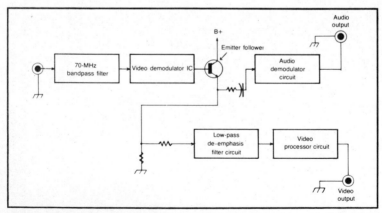

Fig. 6-7. Block diagram of satellite receiver using external downconverter. Input to receiver is 70 MHz intermediate frequency. No downconversion is used in the receiver itself, but the unit is still referred to as a single downconversion type. (Radio-Electronics)

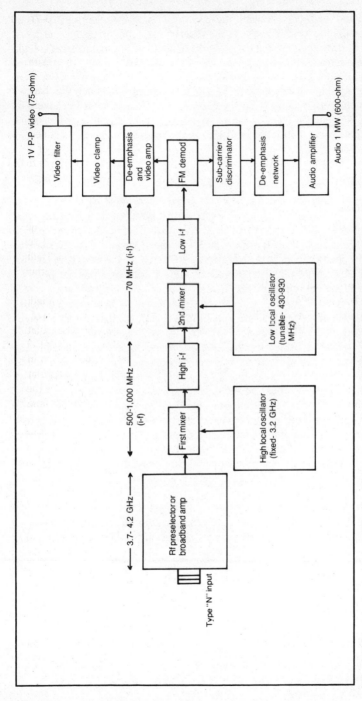

Fig. 6-8. Double conversion receiver. Input is downlink signal from low-noise amplifier. This arrangement is suitable if dish is located near the receiver. Most TVROs today have the downconverter adjacent to or close to the LNA, so receivers get 70 MHz i-f input. (Radio-Electronics)

Satellite receivers have the option of using two different types of demodulators. One of these is the discriminator, a circuit that has been used in FM receivers for many years. The other is a more recent circuit development known as a phase-locked loop or PLL.

Both the PLL type of demodulator and the discriminator have their advantages and disadvantages. The PLL is a more sensitive demodulator, but if the input signal is strong enough, the discriminator is capable of producing quality color pictures. When the input signal approaches its threshold value, a PLL demodulator may be required to produce a viewable picture.

## REMODULATION

Following the demodulator, we have our original baseband signals once again—video and audio. The carrier that is selected for the remodulation process is that of VHF Channel 3 or VHF Channel 4. The frequency band for Channel 3 is 60 MHz to 66 MHz; the video carrier is 61.25 MHz and the sound carrier is 65.75 MHz. For Channel 4, the frequency band is 66-72 MHz; the video carrier is 67.25 MHz and the audio carrier is 71.75 MHz. While the remodulation circuitry is often contained within the satellite receiver, it can also be a separate component. Whether an individual component or associated with a satellite receiver, the user has an option of selecting either VHF Channel 3 or VHF Channel 4 as the carrier frequency to be modulated. Generally, the selected channel is one that is normally not viewed and so is available for this use. However, whichever one of these channels is selected for satellite TV viewing, the tuning control on the front of the TV receiver must be set to correspond.

Once you have set the channel selector control on the TV receiver to either Channel 3 or Channel 4, no further TV receiver control changes will be needed, assuming that picture quality and color are satisfactory. If you get a terrestrial TV broadcast picture on either of these channels, you can expect to get as good a picture (and quite possibly much better) using satellite reception, assuming correct TVRO installation and dish adjustment.

If you are in a fringe area and broadcast TV picture reception for Channel 3 or 4 is poor, this does not mean satellite reception will be bad. Poor reception in a fringe area may not be the fault of the television set, but can be due to weak signals. If you are in a fringe area, and a few channels produce reasonably good pictures, but Channel 3 and 4 do not, you can expect good satellite pictures. However, if the TV set shows poor pictures at all times, even in areas of good signal strength, it is unlikely that satellite TV will improve matters.

## THE NEED FOR A GOOD TV SET

The quality of a TVRO installation can be no better than the quality of the television receiver, and sometimes the receiver is the weakest and the poorest link. If TV picture quality is consistently poor, even for strong or moderately strong TV broadcast signals, it will also be poor for satellite TV. Sometimes, when installing a TVRO, it may be advisable to replace an existing TV set. A black-and-white set can be used but it would negate the purpose of a TVRO.

## THRESHOLD

This is the level of carrier-to-noise needed to supply an adequate signal-to-noise ratio at the output of the demodulator. This output is generally in the order of approximately 46 dB. A picture that is free of sparkles can be assumed to have reached or exceeded the threshold level.

## SIGNAL CLEANUP

Before being presented to the TV set, the video signal must literally be *cleaned up*, a technique that could be called de-emphasis and de-dithering.

The earth station not only transmits the uplink signal but also processes it prior to its long journey to a satellite. One of these processing routines involves pre-emphasis, in which the high-frequency section of the video signal is emphasized. In the satellite receiver, a circuit is used to de-emphasize that part of the video signal, reducing it to overcome the effects of pre-emphasis.

The earth station not only pre-emphasizes the video signal prior to modulation, but also dithers it, or disperses it, at a very low frequency rate, 30 Hz. The purpose of this is to prevent downlink signal interference to earthbound microwave transmissions. Consequently, in the satellite receiver, the video signal is de-dithered, or un-dithered if you prefer. Without the de-dithering circuit of a satellite receiver, the picture on the television screen would flicker at a 30 Hz rate.

## AUTOMATIC FREQUENCY CONTROL

One of the circuits used in the satellite receiver is known as automatic frequency control, abbreviated as AFC. In broadcast band AM and FM receivers, AFC is used to keep the receiver locked in on a station, that is, accurately tuned. By tuning, what is meant is that the receiver's tuning circuits are kept as close to the center of the carrier frequency of the broadcast station as possible. In a television receiver, if the receiver's tuning circuitry should wander, that is, become even slightly detuned, the result would be distortion of both picture and sound.

AFC is also used in the satellite receiver. The 70 MHz intermediate frequency signal has a bandwidth of 36 MHz. The purpose of AFC is to keep the receiver tuned to the center of that band and to hold it there.

## AUDIO SELECTION

Like earthbound television stations, the transponders used in satellites also have subcarriers for the transmission of additional signals. Thus, the second carrier produced by a transponder, known as a subcarrier, is used to carry the audio signal. Thus, the two signals, video and audio, are transmitted simultaneously, but they are always separated by a fixed amount in terms of frequency. The satellite receiver is equipped with a control to let the viewer choose the correct subcarrier for the particular video signal that he is watching.

At the present, there are four basic audio formats used by the various satellites. These include monophonic sound (generally abbreviated as mono but also known as single channel sound), matrix stereo, discrete stereo, and multiplex stereo. All the stereo formats, no matter how derived, are stereophonic (generally abbreviated as stereo but also known as two-channel sound).

While the usual satellite receiver does permit reception of the audio subcarriers, the component is limited to monophonic sound only. Some satellites do supply audio only and this is always stereo.

With some satellite receivers the audio signals are *tunable*. For those who are accustomed to the fixed audio supplied by television receivers this is a new concept. With a fully tunable audio control you can listen to a music program without necessarily watching a picture.

The audio tune control on the front panel of the satellite receiver is used to optimize the sound accompanying a video program or to tune for an audio only program available on some channels. The range of the audio tuning control is from 5.5 MHz to 7.5 MHz.

Note that terrestrial television stations use a single subcarrier only for audio. Consequently television sets for such transmissions do not need to have any controls for tuning in the audio. The audio is always a fixed amount higher in frequency then the video carrier.

However, because a number of subcarrier frequencies are used for audio transmission by the satellites, the satellite receiver requires a control for tuning in the desired audio signal. Like the modulated video signal, the audio requires a demodulator and generally a phase-locked loop (PLL) type is used for this purpose.

## REMOTE CONTROL OF THE DISH

For the ultimate in convenience, the satellite receiver can work with an additional component for remote control of the dish. With its help, the dish can be made to turn to any satellite available in your location. A preferable unit is one that can be programmed to allow return to any previously selected satellite. See Fig. 6-9. Remote control is suitable only for use in connection with a polar-type dish mount.

## FILTERS

A satellite receiver may be equipped with various kinds of filters. The purpose of these filters is to permit passage of the desired signals, and to

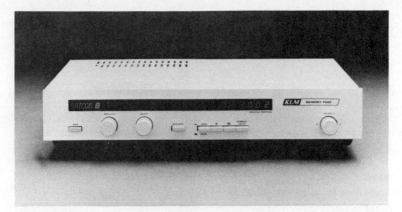

Fig. 6-9. Memory-Trak Motor Control Console. Programs and recalls up to 50 satellite positions, automatically. Features digital satellite and position readout panel, manual override, polarity control (Mag/Luly type), 5 day memory retention if power fails. (KLM Electronics, Inc.)

bypass or eliminate any unwanted signals. There are numerous types of filters, but the three basic ones are low pass, high pass, and bandpass. A low pass-filter is one designed to pass all frequencies up to a selected frequency and to suppress or bypass all others above the selected frequency. A high-pass filter is exactly the opposite. It is intended to pass all frequencies above a selected frequency and to suppress or bypass all frequencies below the selected frequency. A bandpass filter will pass all frequencies between two selected frequencies—a low frequency and a higher frequency. All frequencies above or below the selected frequencies are rejected.

The i-f section of a satellite receiver can be looked on as a bandpass filter. Another example is the use of a narrow/wide bandwidth control on the satellite receiver, if it is so equipped. Reception of weak or noisy signals can be substantially improved by adjusting the control to its *narrow* setting.

## SAW FILTERS

The intermediate frequency received at the input of the satellite from the downconverter is 70 MHz but since this signal is frequency modulated, it consists not of a single frequency but a group that has a maximum bandwidth of 36 MHz. This is an extremely wide band and so it is difficult for the intermediate frequency stages to handle. Ideally, the i-f stages should accept only those frequencies that are within the i-f bandpass and reject all other signals, but the wider the bandpass the more difficult this is to do. Signals outside the pass band, but which may not be rejected by the i-f, could be noise or interfering signals.

As an aid in obtaining a well-defined bandpass, a SAW filter (surface

155

acoustic wave) is sometimes used in better grade receivers. However, a SAW filter has a relatively high inherent insertion loss of about 30 dB and so it is sometimes necessary to use one, and sometimes two amplifiers to recover signal strength. Additional amplifiers, when used, are placed in line with the SAW filter.

Surface acoustic wave filters are small, rugged and very well shielded. The use of a SAW filter, plus additional amplifiers increases the manufacturing cost of a satellite receiver, but it does result in less noisy, more interference free pictures. An advantage of the SAW filter is that it completely eliminates the need for fine tuning of the filter, or adjustment.

SAW filters are made on a variety of piezoelectric substrates, the two most common being lithium niobate ($LiNbO_3$) and ST-quartz. Quartz has an excellent temperature stability over a wide range, while lithium niobate exhibits excellent electromagnetic acoustic coupling.

## AUTOMATIC GAIN CONTROL

Automatic gain control (agc) consists of one or more circuits that boost or decrease gain so as to maintain output at a fairly constant level.

## CONNECTING THE REMODULATOR TO THE TV RECEIVER

Whether the remodulator is a separate component or is part of the satellite receiver, its output is the composite video and sound signals modulated onto a Channel 3 or Channel 4 carrier. The remaining step in a TVRO system is to connect that output to the antenna terminals of the television receiver.

The television receiver has terminals for VHF and UHF reception. As far as the TVRO is concerned, the UHF terminals can be ignored. Since the VHF terminals will already be occupied by a lead-in from the antenna for broadcast TV reception, there is now the problem of also connecting the coaxial cable from the remodulator. This can be done, as shown in Fig. 6-10

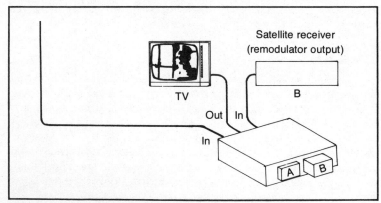

Fig. 6-10. Simple A/B switcher for broadcast or satellite TV.

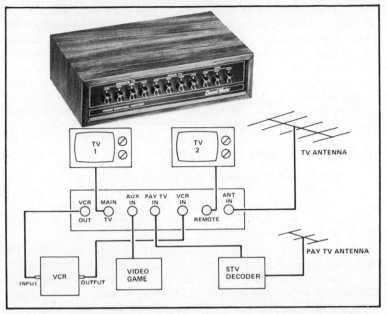

Fig. 6-11. Unit makes selection of video source easy. It permits connecting as many as four different signal sources to two separate TV sets. Uses pushbuttons to switch from one piece of video equipment to another. It has a frequency range of 5 MHz to 890 MHz and isolation of more than 60 dB between signal sources for VHF; 40 dB for UHF. Does not need to be connected to power line. (Channel Master, Div. Avnet, Inc.)

by using a switcher for program selection. This switcher allows the selection of either broadcast or satellite TV programs.

The disadvantage of the unit shown in Fig. 6-10 is that it permits the selection of just two program sources. A better arrangement is the component illustrated in Fig. 6-11, which allows a choice of as many as four video program sources.

## SATELLITE RECEIVER VIDEO OUTPUT

The baseband signal output of the satellite receiver consisting of video and audio is connected to a jack on the rear of the component. This output is comparatively low frequency and so it can be handled by a type of coaxial cable identified as RG-59. The output impedance at the video output is 75 ohms and so is the characteristic impedance of the cable, hence impedance matching is automatic.

The video signal can then be fed to an external remodulator (often more simply known as a modulator) for subsequent delivery to the antenna terminals of a TV receiver. Alternatively, the signals can be delivered to a videocassette recorder. Such a unit contains a built-in remodulator, so in this case such a component will not be required.

A control on the rear of the satellite receiver may be available for adjusting the strength of the baseband signals. This output, if checked with a test instrument, should be 1.25 volts, peak-to-peak. One way of setting this control without the use of test instruments is to have the satellite receiver put a picture on the TV screen and then to adjust the control for the best picture. If the picture looks weak and washed out, turn the video control for more signal output. If the picture looks very strong, but has a tendency to tear or skew, the video driving voltage is too high and the video level control needs to be turned down.

## THE STEREO PROCESSOR

Ultimately, the video and audio signals returned to earth by satellite will need to be delivered to some component that will display the pictures and reproduce the sound. No matter how fine the quality of the picture or how good the sound may be, these can be no better than the quality of the TV receiver.

As far as audio is concerned, the only word that can be used to describe the sound supplied by the average TV set is pathetic. The sound is monophonic, the speakers are tiny, the enclosure (the TV console) is not designed for sound reproduction, and the kindest thing that can be said about the audio amplifier in the TV set is that it is minimal.

In short, the usual TV set is wholly inadequate for reproducing sound. To be able to enjoy all the formats of stereo music and all the sound tracks available on satellite TV, it is possible to use an add-on outboard component, one that can be connected to and work with the satellite TV receiver. This component, known as a stereo processor, and pictured in Fig. 6-12 is an indoor unit and can be positioned adjacent to, above, or below the satellite receiver.

A stereo processor enables the user to avoid TV *dead end* sound in two ways. The processor contains its own audio power amplifier, and since this will be a stereo unit, it will be able to deliver two-channel sound. Further, the audio power amplifier section in the stereo processor will be able to drive a pair of high-efficiency, quality loudspeakers, directly. Alternatively, it is possible to deliver the stereo processor's two-channel sound output to the input of a high-fidelity sound system.

Because the stereo processor has outputs for a speaker system (unlike a TV set it does not have built-in speakers), the external speakers can be positioned properly for best stereo effect.

Some of the transponders handle stereo programs; others mono. The decision as to mono or stereo transmission is not a function of the satellite, but depends on the uplink signal. Publications listing satellite programs sometimes indicate whether the audio is stereo.

It is possible to use the stereo processor to obtain any one of the four different types of sound listed earlier: mono, matrix stereo, discrete stereo, or multiplex stereo.

Even if a satellite transmits mono sound only, it is possible to convert

Fig. 6-12. Stereo Processor can be used to drive hi/fi system. Unit has separate A and B subcarrier tuning controls and center-tune meters. This component has narrow and wide bandwidth selection, dynamic noise reduction, Anik filtering, expansion switch, LED multiplex (MPX) indicator, video and mono audio output jacks. Optional amplifier section includes volume control, headphone jack, left and right speaker terminals and bass boost switch on the rear panel. (KLM Electronics, Inc.)

it into a *synthetic* type of stereo by using a stereo synthesizer, an accessory add-on unit, in conjunction with a hi/fi system.

The stereo processor is equipped with jacks for headphone use, permitting stereo listening in privacy, and with minimum distortion. It is minimum distortion since the headphones are connected at a point prior to the TV set. Finally, it is only with the help of a stereo processor that it is possible to enjoy stereo sound via satellite.

## Stereo Processor as a Receiver

In a way, the satellite stereo processor is a receiver, but not so in the usual sense. It is actually an adjunct to the satellite receiver itself. The stereo processor obtains the sound signal from the satellite receiver and since that sound signal is FM, the stereo processor can be looked on as a type of FM receiver.

A local oscillator in the processor heterodynes or beats with the signal obtained from the satellite receiver, producing an intermediate frequency of 10.7 MHz. Note that this intermediate frequency is the same as that used in ordinary FM receivers that pick up earthbound FM signals.

The tuning range of the stereo processor is 5.5 MHz to 8.0 MHz, enabling the unit to select any of the available subcarriers. In some units, the bandwidth of the intermediate frequency is made bandwidth selectable—that is, the bandwidth of the i-f signal can be either 150 kHz or 500 kHz. A narrow bandwidth means greater gain, but some of the musical information can be lost. A wider bandwidth means less gain, but it also means the entire signal is passed. A narrow bandwidth may be desirable if the audio signal is accompanied by excessive noise or if the audio signal is somewhat weak.

# Chapter 7

# How to Install a TVRO System

Before you can install a TVRO system, you will need to have two bits of information, one of which may appear strange. You will need to know where the satellites are, which is fairly obvious. You will also need to know where you are, more specifically, where you are on earth. That isn't as difficult as it sounds and it starts with our method of dividing the earth into sections using lines of latitude and longitude.

## LATITUDE

Latitude, sometimes referred to as lines of latitude or parallels of latitude, are imaginary lines that encircle the earth in true east-west directions, as shown in Fig. 7-1. The reference line used for lines of latitude is the equator, which is designated as latitude 0 degrees. The true North Pole is latitude 90 degrees and so is the true South Pole.

To distinguish one from the other in terms of latitude, lines of latitude north of the equator are accompanied by the letter N, for north, and those south of the equator with the letter S, for south. Thus, the true North Pole is 90 degrees N (or north latitude) and the true South Pole is 90 degrees S (or south latitude). Lines of latitude are sometimes called parallels since they are parallel to each other and to the equator.

The representation of lines of latitude depends on whether they are depicted on a three-dimensional object, such as a globe, or on a flat surface, such as a map, also on the particular projection used in making the map. On a globe, lines of latitude are represented by circles. Of these circles, the equator is the largest. The circles become progressively smaller as the parallels get closer to the poles. The poles and the equator represent the extremes in terms of degrees of lines of latitude. A line of latitude, such as 40 degrees north would be a line encircling the globe parallel to the equator.

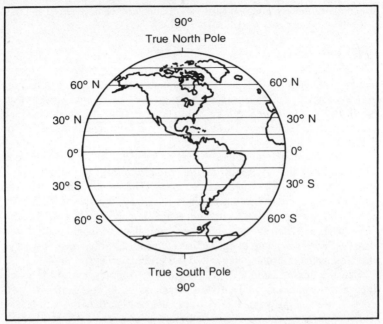

Fig. 7-1. Parallels of latitude. One degree of latitude is equivalent to about 69 statute miles. The equator is the reference and is zero degrees latitude. Lines of latitude above the equator are north; below the equator are south. The true North Pole is 90° north latitude; the true South Pole is 90° south latitude.

Such a line would pass east to west completely through the continental U.S.

Each degree of latitude is approximately 69 statute miles. Consequently, a line of latitude of 40 degrees N would be 69 × 40 = 2,760 miles. This means that all points on that line of latitude would be 2,760 miles north of the equator. Since the true North Pole is 90 degrees, that same line of latitude (40 degrees) would be 90 − 40 = 50 degrees. 50 × 69 = 3,450 and so a line of latitude of 40 degrees is 3,450 miles south of the true North Pole. These figures are approximate, since the earth isn't a perfect sphere but has something of a bulge at the equator and is somewhat flattened at the poles.

While lines of latitude are measured in terms of degrees, each degree can be subdivided into minutes. A degree contains 60 minutes. Each minute can be further subdivided into 60 equal parts called seconds. These subdivisions help make the determination of a location much more precise, but in making such determinations the use of lines of latitude in degrees only is adequate.

A line of latitude is precisely that, a line, and so while it is helpful in determining location, it cannot be used alone. Thus, the city of Tel Aviv in Israel and Nanking in China are both on latitude 32 degrees N, but these cities are about halfway around the world from each other. To help fix the specific location of a point on earth, another reference is required and this reference is known as longitude.

## LONGITUDE

Since lines of latitude can only determine north/south position, another reference is needed and that is supplied by lines of longitude (Fig. 7-2) also known as meridians. Meridians are lines that extend from the true North Pole to the true South Pole. Just as the equator is the starting or reference line for lines of latitude, so too is a reference line needed for longitude.

The line of longitude selected for reference is known as the Greenwich meridian. This line extends from the true North Pole to the true South Pole, passing through Greenwich, England. This was originally selected as the reference meridian since it passed through the site of the former royal observatory in Greenwich, but which is now part of London. The Greenwich meridian is the reference for all lines of longitude and is 0 degrees. Each meridian, then, is a line of longitude, either east or west of the Greenwich meridian, also called the prime meridian.

Just as lines of latitude divide the earth horizontally, meridians divide the earth longitudinally. Thus, the meridians of 90 degrees east or 90 degrees west of the prime meridian are each one-fourth of the world away from the prime meridian. The meridian of 180 degrees east or west of the prime meridian is half-way around the world from it.

Meridians are designated in degrees and these are accompanied by the letters E or W to indicate direction from the Greenwich meridian. There is one exception and this is the meridian at 180 degrees. Since it is equally distant east or west from the prime meridian, it is often designated on maps as E 180 W.

Lines of longitude get closer to each other in their approach to either pole. At the equator, each degree of longitude is approximately 69 miles. Note that this is the same separation as that for lines of latitude, an expected result since the earth is so close to being a sphere. However, for lines of longitude this is the maximum separation. The closer the approach to either of the true poles, the shorter the separation. Halfway between the equator and either the true North or South Pole, one degree of longitude is approximately 49 miles. At either the true North or South Pole, theoretically, at least, all lines of longitude should intersect at a point.

Just as the location of a place on earth cannot be determined by latitude alone, so too is it impossible to find a place by longitude alone. Thus, Cairo, Egypt and Durban, South Africa, are on the same line of longitude, 31 degrees E, but both are thousands of miles apart. By indicating the latitude and longitude of a city, for example, its location on the surface of the earth can be determined.

## THE DISH AND LINES OF LATITUDE/LONGITUDE

To be able to find one or more satellites, it is essential to know exactly

162

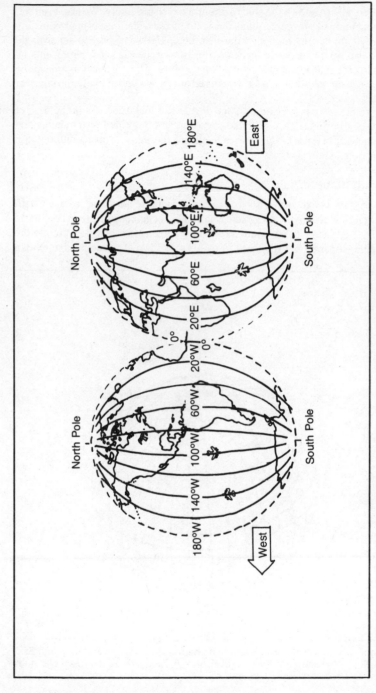

Fig. 7-2. Lines of longitude, also known as meridians, they converge at the true North and South Poles.

163

where we are so as to be able to orient the dish precisely. We are, in effect, trying to *shoot* a target so small it is invisible, at a distance of approximately 22,300 miles. Any motion of the dish, up or sideways sweeps out an angle that can be measured in inches at the face of the dish, but represents an angle of thousands of miles out at the satellite. Thus, a small motion of the dish can mean either a much weakened signal, or, more likely, no signal at all.

The earth is divided into 360 degrees of longitude. As in the case of degrees latitude, each of these degrees can be divided into minutes and seconds. For each 1.5 degrees of longitude there is one hour's difference in sun time.

## MAGNETIC DEVIATION

North is expressed in two ways: as magnetic north and as true north. Magnetic north is the direction supplied by a compass pointer. True north (Fig. 7-3) is the north-south axis of the earth. Magnetic deviation is the angular separation, in degrees, between these two. The compass supplies a

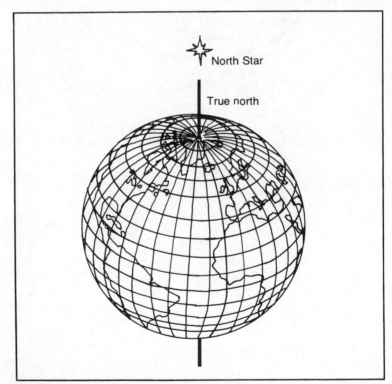

Fig. 7-3. True north is at one end of the earth's axis and points to the North Star. Unlike magnetic north, true north is always fixed in position. Magnetic north moves several minutes a year.

magnetic indication; true north is a geodetic indication. Similarly, geodetic south is true south. Magnetic deviation (also called variation or declination) is the angle between true north and magnetic north. The magnetic declination in the U.S. ranges from about 25 degrees east in the northwestern portion of the country to about 15 degrees west in the northeastern portion of the country. There is a gradual change in the magnetic declination in any one locality from year to year. This change can be predicted from past records. In some parts of the U.S. this amounts to as much as 4 minutes annually. Since there are 60 minutes in a degree, the maximum change of one degree requires 15 years. In an area where the change in magnetic declination is only 3 minutes annually, it will take 20 years for a one-degree change.

## COMPASS HEADING

This is the magnetic direction supplied by a compass. Calibrated in degrees, it indicates the direction of magnetic north.

## COMPASS DEVIATION CORRECTION

True south follows the longitudinal lines on a map or globe. A compass will deviate from true south by as much as 22 degrees in continental U.S. The map shown in Fig. 7-4 can be used to determine the average magnetic correction needed for various areas. For instance, in central Oregon, true south will be found 20 degrees to the east of magnetic or compass south. In Maine, true south will be found 20 degrees to the west of magnetic or compass south.

As shown on the map, compass deviations aren't straight lines, but curve somewhat. Note also that in some areas there is no compass deviation and that magnetic north is synonymous with true north.

## COMPASS ROSE

Most navigation charts show a compass rose which indicates the amount of variation (magnetic declination) for a given locality. Such a compass rose, shown in Fig. 7-5, consists of a double graduated circle. The outer one is marked in degrees and the inner one in compass points. Zero degrees on the outer circle is always oriented to true north. The inner circle is oriented to magnetic north and it shows the variation for the locality of the chart and for the year designated at the center of the rose. In this illustration, magnetic north is exactly 13 degrees west of true north for the year 1984. Also, the legend at the center of the compass rose notes an annual increase of two minutes. If the chart is read in 1990, then the total increase in variation is 14, that is the number of years multiplied by two minutes. Thus, in 1990 the total magnetic declination would be 13 degrees 14 minutes.

To convert a bearing obtained by a compass to a true bearing (a bearing with respect to true north) the following procedure is followed: when the

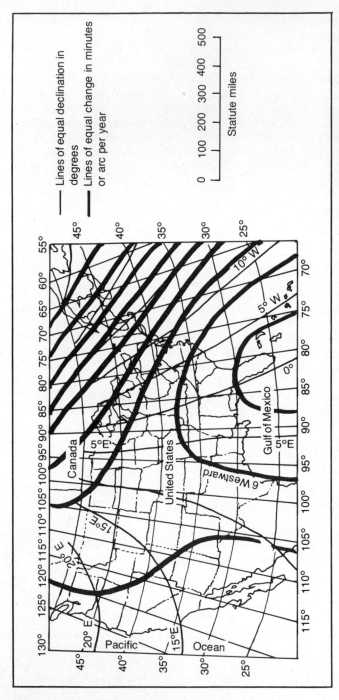

Fig. 7-4. True south follows the longitude (vertical) lines on a map or globe. A magnetic compass will deviate from true south by as much as 22° in continental U.S. This map can be used to determine the average magnetic correction needed in your area. For example, in central Oregon, true south will be found 20° to the east of magnetic or compass south. In Maine, true south will be found 20° to the west of magnetic or compass south. (M/A Com Video Satellite, Inc.)

—— Lines of equal declination in degrees

—— Lines of equal change in minutes or arc per year

0   100  200  300  400  500

Statute miles

166

variation is westerly, then the total variation is subtracted from the magnetic bearing to get the true bearing. When the variation is easterly, the total variation is added to the magnetic bearing to get the true bearing.

## PRECAUTIONS WHEN USING A COMPASS

It is important to remember that the compass will be affected by nearby iron or steel objects, including the dish mount. Consequently it is preferable to make a site survey without the polar mount, usually made of steel, in the immediate vicinity. Once the TVRO is functioning, the accuracy of all the settings, including declination, deviation, and true south can be verified by clean, on-target reception of satellites at the far ends and the middle of the satellite chain.

## SELECTING THE SITE

The first step in installing a dish is to make a selection of the site. The site must have an acceptable *look angle*, that is, it must have a clear, unobstructed view of the satellite or satellites to be used for signal recep-

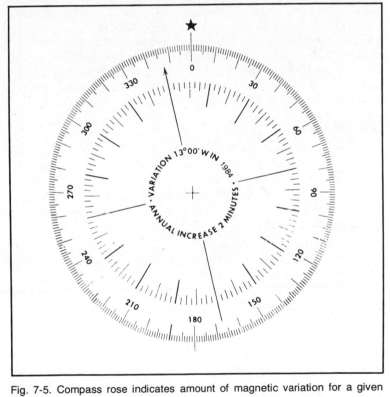

Fig. 7-5. Compass rose indicates amount of magnetic variation for a given locality.

tion. Once a possible site with an acceptable view is selected, there are some additional factors to consider.

One of these is the possibility of micro-link interference. Microwave signals, having the same frequencies as the satellites, are often used commercially (by the Bell system, for example) as terrestrial microwave links. Generally, if a microwave tower is visible from the site, you can expect trouble.

## POWER LINES

It is inadvisable to install a dish close to high-voltage, overhead power lines. Such lines generate a large amount of noise and under certain conditions can totally override the satellite television signal. Some dealers who handle TVRO systems have portable setups that will help determine the suitability of a site.

## UNDERGROUND CONSTRUCTION

Before permitting any digging for a base for the mount, make certain there are no existing underground installations. Telephone lines, power lines, storm drains, and other installations are often put underground and their exact location is not generally known to everyone.

Information concerning possible underground installations should be obtained from the property owner or other appropriate agency such as the telephone company, electric power company, or city planning office. If a site cannot be proven clear of underground obstructions, a survey with a metal detector may be needed before digging begins.

## SOIL CONDITIONS

When selecting a TVRO site, consider soil conditions. Soil varies widely in composition and load bearing capacity. Compacted gravel provides a much firmer base than does soft clay. In many cases an 18-inch thick solid concrete slab will provide a stable base.

## LOCAL BUILDING CODES

At the present time no known restrictions exist concerning the placement of TVRO equipment. As installations become more common however, some communities are sure to regulate their placement.

## FOUNDATION SELECTION AND CONSTRUCTION

A TVRO system usually requires a large parabolic reflector type dish that must be maintained in a fixed position while being subjected to wind loading. The forces imposed on a dish by a survival wind load are significant due to the reflector's large surface area. Because the center of the reflector may be many feet above the ground, the foundation must not only resist horizontal drag forces but large overturning forces as well. The foundation must safely transfer these forces into the ground, providing a sturdy base that will keep the dish movements within minimum allowable limits.

A satellite dish is designed to withstand three kinds of loads that can reasonably be expected to occur during its useful life. The weight of the dish will always have to be supported and this is known as its dead load. The dish will be exposed to all types of local weather conditions and during winter months there may be the formation of a layer of ice. This additional weight of ice is called the ice load. The dish will also be subjected to forces generated by the wind, called wind load.

The number of satellites whose programs you will be able to watch and the quality of reception will be determined by the site location. In some instances, the cooperation of neighbors may be needed since they will always be concerned with possible changes in the value of their property. As far as your own property is concerned, the installation of a dish should have the effect of increasing its value, much as a swimming pool or other substantial improvement would.

As a start, consider that any good sized dish, such as one having a diameter of 10 feet or more, will be ground mounted. This means you must have a sufficiently clear space around the dish. Typically, a dish having a diameter of 10 or 11 feet would require a physical space 12 feet square. This is not the same as 12 square feet. 12 feet square means a geometric figure each of whose sides is 12 feet. The total area of such a base is $12 \times 12 = 144$ square feet. A smaller area can be used if the dish has adequate clearance both horizontally and vertically.

The dish will not only require sufficient base area, but has certain height requirements as well. For this example, allow a space of 12 feet vertically from the bottom of the dish support to the top of the dish.

Ultimately, the signal picked up by the dish must find its way to the television receiver (or receivers) in the home. Ideally, the dish should be located as close to the in-home TV as possible, but this is not always practical. A maximum distance would be about 300 feet. Shorter distances are desirable, since the connecting coaxial cable between the indoor and outdoor components provides signal loss. The shorter the cable, the lower this loss.

The face of the dish must have a clear, unobstructed line of sight view of the satellites you wish to receive. If this view is partially or totally blocked by trees, buildings, terrain, etc., reception will not be acceptable. If necessary, trees or other natural obstacles should be trimmed or removed.

It may not be necessary to cut down trees and it may only be necessary to trim branches. However, branches and leaves do grow back, so when trimming is called for do enough of it so you do not have to repeat this action every spring.

The dish not only points upward into the southwestern sky but for reception of all the satellites it will be necessary for the dish to swing through an arc of about 90 degrees, also changing its elevation as it does so.

the site is suitable. This is only for one satellite though. If you want to be This is sometimes referred to as the field of view of the dish. If you want to view all the satellites, then the field of view will be maximum. This may not be desirable if your interest is in the programs supplied by only one or two satellites that are adjacent in orbit. While this means reception is limited, it also makes installation that much easier. If your site is congested with obstacles you may need to compromise, choosing a field of view that will give you a clear line of sight to the satellites of your choice.

## Microwave Telephone Interference

Telephone companies use the same microwave frequencies as those used by the satellites whose signals you are planning to receive. If you are located in a metropolitan area, it is advisable to learn if your home is in the path of telephone microwave transmission and reception. Such transmissions can interfere with you as much as you would interfere with them. Generally, you will not find this to be a problem, since microwave telephone transmissions are beamed, using dishes, from one location to the next. This step is to make sure you aren't directly in the path of such a beam.

## THE SITE TESTER

Before any dish installation can be considered, it is essential to make a site survey. The survey will tell you just which satellites your dish will be able to see. You cannot simply stand in a particular spot and look up into the southern sky though.

To determine the suitability of a site, it is necessary to use a site tester. There are a number of these, some of which are home-made

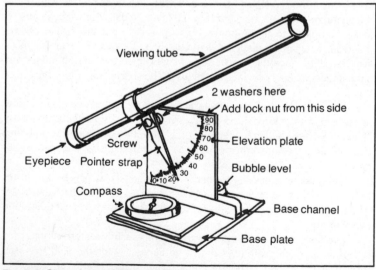

Fig. 7-6. Site selector device. (KLM Electronics, Inc.)

devices, others somewhat more sophisticated. Figure 7-6 is an illustration of a commercially available site tester. It resembles a miniature telescope (although it has no lenses) and is equipped with an elevation plate calibrated in degrees, a bubble glass to help keep the unit perfectly horizontal, and a compass.

While the distance of a satellite from the earth is often supplied as 22,300 miles, this is the distance from a satellite to a point on the equator directly below it. However, TVROs are not only located well north and south of the equator, but may be placed so that the line of sight is a long slant line. Thus, the minimum distance shown in Fig. 7-7 is 23,321 miles while the maximum is 25,383 miles.

To be able to use the site tester, you will need to know the azimuth and elevation headings for each of the satellites from your location. There are commercial services that will supply you with a computer printout or it will be supplied by the manufacturer of the dish. The printout (Fig. 7-7) lists the azimuth and elevation headings for each of the satellites at your location. The printout figures are given in both true and compass headings.

It will be necessary to position the site tester about six feet above the ground and for this you will need a ladder. Since the compass is sensitive to ferromagnetic materials, such as iron or steel, it will be necessary to remove all of these, since they will affect the movement of the compass needle. If you have a vehicle in the vicinity move it as far away as you can. The same is true of personal objects that may contain ferrous materials, such as a metallic belt buckle. You can easily check if any personal object is affecting the compass since the compass pointer will move every time you bring the object close to it.

Select the compass heading for the first of the satellites you have in mind. If you have no particular preference you can start with the compass heading for Satcom 1. This is a popular entertainment satellite at the western end of the chain. Keep in mind that what you are looking at is the compass heading and not the true south heading. Now rotate the dial on the compass until the selected compass heading is aligned with the line marked *read bearing here*. The illustration shown in Fig. 7-8 indicates a bearing of 120 degrees.

Rotate the site tester until the compass needle points to north. At this time, tilt the viewing tube until the pointer indicates the elevation heading from the computer printout for this satellite (Satcom 1). The illustration (Fig. 7-9) shows an elevation angle of 35 degrees. The elevation you read on the site tester can be more or less than this, depending on your location.

The site tester is equipped with a bubble glass (Fig. 7-10). The glass has an inner circle or ring. Be certain the bubble is inside this ring. The purpose of the bubble glass is to make sure the site tester is absolutely parallel with the surface of the earth. If it isn't, it will be necessary to put paper or cardboard shims under the site tester.

Now take a look through the viewing tube. If you get a clear view of the sky, if there are no obstructions in the way, you have the first indication that

```
CORRECTION FOR TRUE NORTH   11 WEST

EARTH STATION COORDINATES - DEGREES,MINUTES,SECONDS
LONGITUDE  74   22   30      W
LATITUDE   40    0    0      N

AZIMUTH FIGURES ARE GIVEN IN 'TRUE' AND 'COMPASS' READINGS
USE THE 'COMPASS' READINGS FOR SITE TESTING

SATELLITE      LONG.    AZIMUTH    ELEVATION  DISTANCE   'COMPASS' AZIMUTH
WESTAR 1/2      79       187.173    43.4877    23321      198.173
SATCOM F4       83       193.278    42.8802    23348.4    204.278
COMSTAR D3      87       199.211    41.9247    23392.2    210.211
WESTAR 3        91       204.916    40.6447    23451.9    215.916
COMSTAR D1/D2   95       210.35     39.0699    23527.3    221.35
WESTAR 4        99       215.492    37.231     23617.8    226.492
ANIK D1         104      221.498    34.6123    23751.2    232.498
ANIK B          109      227.049    31.6945    23905.8    238.049
ANIK A3         114      232.178    28.5354    24080.1    243.178
SATCOM F2       119      236.926    25.1842    24272.3    247.926
WESTAR 5        123      240.482    22.3944    24437.8    251.482
COMSTAR D4      127      243.849    19.5282    24612.5    254.849
SATCOM F3       131      247.05     16.6027    24795.5    258.05
SATCOM F1       135      250.109    13.6318    24985.7    261.109
SATCOM F5       143      255.879    7.60346    25383.2    266.879
```

Fig. 7-7. Computer printout supplying azimuth/elevation coordinates for central New Jersey.

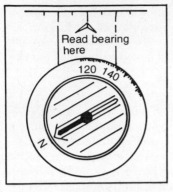

Fig. 7-8. When using the site selector, rotate the dial on the compass until the selected compass heading is aligned with the line marked *read bearing here.*

able to receive all the satellites, repeat the site testing checks on a satellite at the other end of the range. If that is clear, then you know the outer limits of satellite reception are acceptable. Make a number of checks on satellites between these two to make sure the region between them is clear. If it is, then the site you have selected can be used.

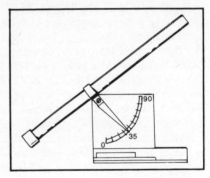

Fig. 7-9. Site tester measures angle of elevation. (KLM Electronics, Inc.)

The first site that is chosen may not always be the best one for satellite viewing. You may have picked it since it supplies ready access or is the spot closest to the house. If it doesn't supply a clear view of the satellites you have chosen it isn't satisfactory. You can still use that site, however, if it is possible for you to remove the obstructions.

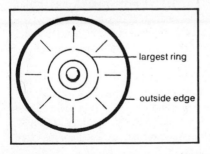

Fig. 7-10. Bubble glass is used to make sure site tester is perfectly horizontal. (KLM Electronics, Inc.)

If this cannot be done, make a second site selection. Because you have now acquired some experience with the site selector, you will find yourself able to do this much more rapidly. Because you now understand more fully the objective, you will be able to appraise possible site locations more readily.

## SITE FACTORS

A few site factors were discussed earlier, but these were in connection with ground suitability. There are other points to consider. Installing a ground supported dish isn't as simple as putting up a television antenna. There may be local codes or legal hindrances, such as deed covenants, that are restrictive. As a first step, check to learn if there are any other dishes in your neighborhood. One way of doing this is to ask your mailman. Get the name, address and telephone number of the dish owner. Call and identify yourself and try to learn just what difficulties, if any, the owner of the dish had. The problems may or may not be applicable in your case. If, for example, you need to obtain a variance from your township, you can learn just how to go about getting it, and in this way save yourself considerable time with your local bureaucracy. Further, the existence of an operating dish in your locality establishes a precedent, making it easier for you to install one also.

If you rent or lease your property you should obtain permission for the installation of a dish. If you have an attorney, it would be desirable to have him check the legality of dish installation. Some leases indicate that alterations made to a property become the property of the owner when the lessee decides to move. If that is the case with your lease, it may need to be modified so as to protect your investment. If the dish is supported on a poured concrete base you may not be able to move that base when you move to another location but there is no reason why you cannot take the dish with you, plus the connecting coaxial cable between the outside equipment and the house.

## MANUFACTURER'S INSTALLATION

Some manufacturers have dealers trained to make site installations in case you prefer not to be a *do-it-yourselfer*. There are various advantages to making use of such a service. Having made dish installations previously, the technicians employed by a TVRO dealer can make a quick appraisal of the site, determine the best spot for the location of the dish, assemble the dish and its supports, install all the other components, and do so quickly. They will also be equipped to cut coax to its correct length and to attach the necessary connectors.

If you plan to have the dealer's technicians do the installation for you, it is possible to expedite matters by making a preliminary site survey. This is simply an overview sketch such as the one shown in Fig. 7-11. The sketch is made on a sheet of equally spaced horizontal and vertical lines known by the imposing name of Cartesian coordinate graph paper. The paper is often

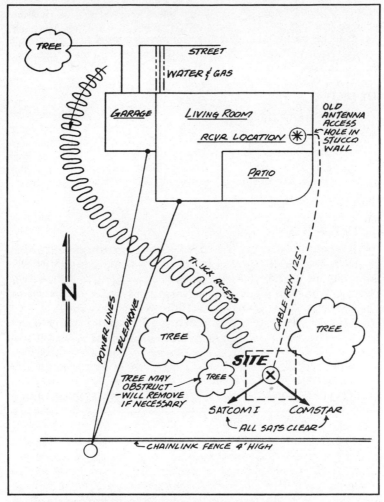

Fig. 7-11. Site overview sketch. (KLM Electronics, Inc.)

supplied by the manufacturer of your satellite dish, but it is also available in art and stationery stores.

All you will need for making the sketch is a ruler and a pencil. Using the site tester, make an estimate of where you think the dish should be located. Measure the distance from the dish to the entry point in the home for the run of coaxial cable. If there is a path that can be followed by a truck to the dish site, indicate this as well.

Note the large letter N shown in the sample sketch. This is magnetic north as indicated by the compass on the site tester. The scale of the sketch is such that each square represents approximately five feet. If there are any trees that may possibly interfere with a clear view, indicate on the sketch whether it will be possible to remove these entirely or if they can be trimmed.

Include major objects near or around the proposed site. This would include trees, shrubs, building poles, power lines, etc. Finally, draw two lines to indicate the outer limits of dish travel by showing two satellites, one to the left and the other to the right.

## THE SITE DATA QUESTIONNAIRE

The proper installation of the dish is basic to the operation of a TVRO. In some instances, it may simply not be possible to install a dish. In an effort to help you make a correct determination, the manufacturer or his dealer may ask a number of questions. A representative site data questionnaire is shown in Fig. 7-12.

## FOUNDATION TYPES

Basically, there are two types of foundations, although these may have a number of variations. One of these is the slab type and consists of a single monolithic slab of reinforced concrete as indicated in Fig. 7-13. The main advantage of the slab type foundation is that it distributes the forces from antenna weight and the various types of loading over a large area. This type of foundation may be required where the soil is very loose or where drainage is a problem. The slab foundation does not depend on soil cohesion to resist the overturning forces due to wind loading. These forces are resisted by the weight of the concrete itself. There are several ways of constructing slab mounts (Fig. 7-14).

Another type of foundation is the drilled pier, and is illustrated in Fig. 7-15. The pier forms a footing that resists movement in any direction as a solid unit.

The selection of a foundation type is often influenced by special problems at the site. If the site is extremely rocky or solid rock, pier holes cannot be drilled with lightweight equipment. In this case a slab foundation may be placed directly on the rock. A slab can also be used in places where the soil is so loose that pier holes would collapse before concrete can be poured.

To prevent foundation upheaval from frost expansion of the soil, the bottom of the foundation should extend below the level of maximum frost penetration. Pier foundations are often used for this reason at sites where excessive frost penetration would require a very deep excavation for a slab.

Even though you will be using concrete as a support for the dish, the ground around and under that base should be firm. If the water ground table is unusually high, that is, very close to the surface, any heavy rain will make the ground beneath the concrete base too soft to support it properly. Areas subject to flooding or heavy runoff aren't suitable.

The selected site must be accessible for assembly and installation. A motorized cement mixer is desirable since mixing cement manually for the

1. Estimate length of cable run between dish antenna and receiver inside your home. Include 6 feet of cabling inside dish and length needed inside your home.

2. Describe picture quality, if any, on channel 3 and 4 of your TV with antenna and/or cables disconnected.

    Channel 3 _____

    Channel 4 _____

3. Which satellites (if any) cannot be received due to obstructions?

_____

4. Please list any questions or comments you have about the selection or suitability of your site or the overall installation.

☐ Yes, I want to install the in-ground 5' X 5' base before installation. Please mail me the necessary instructions.

☐ 16' dish model, for best reception in fringe areas, if recommended in letter accompanying computer printout. (no additional charge)

Special shipping/delivery instructions: _____

_____

_____

Name _____

Address _____
                         (No PO Boxes, please)

Home Phones _____
                Area

Hours you can be reached _____
                            (if consultation is needed)

Signature ✔ _____

If you decline to order, please give reason:

☐ No appropriate site available

☐ Other, please explain _____

_____

_____

Fig. 7-12. Site data questionnaire. (KLM Electronics, Inc.)

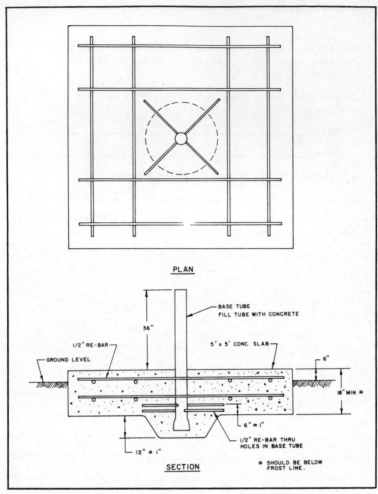

PLAN

BASE TUBE
FILL TUBE WITH CONCRETE

36"

1/2" RE-BAR

5' x 5' CONC. SLAB

GROUND LEVEL

6"

18" MIN *

6" ± 1"

1/2" RE-BAR THRU
HOLES IN BASE TUBE

12" ± 1"

SECTION

* SHOULD BE BELOW
FROST LINE.

Fig. 7-13. Plan and section views of slab installation for polar mount. (Channel Master, Div. Avnet, Inc.)

dish is hard work. There are two types of motorized mixers—gasoline and electrically operated. The electrically operated needs access to a 117-volt ac power outlet.

After pouring the cement, let it cure for a minimum of three days before continuing. On the day following the pouring of the cement, sprinkle it with water and do so several or more times each day. Do not use a forceful jet: a light spray of water is what is needed. If there are weather reports of rain, cover the surface of the cement with plastic, held in place around the edges of the base, using stones or bricks. Do not walk on the cement for at least three days after pouring, even though the surface may seem to be

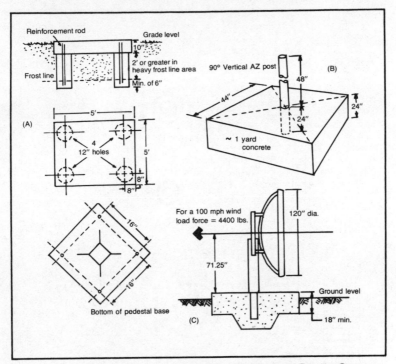

Fig. 7-14. Various suggestions for slab mounts. (A) Winegard Satellite Systems; (B) KLM Electronics, Inc.; (C) Channel Master. (Div. Avnet, Inc.)

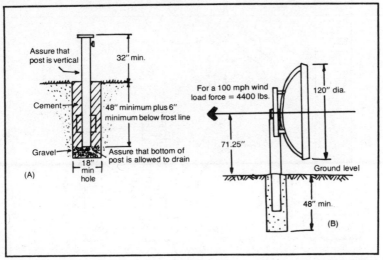

Fig. 7-15. Various pier suggestions for post mounts. (A) Winegard Satellite Systems; (B) Channel Master. (Div. Avnet, Inc.)

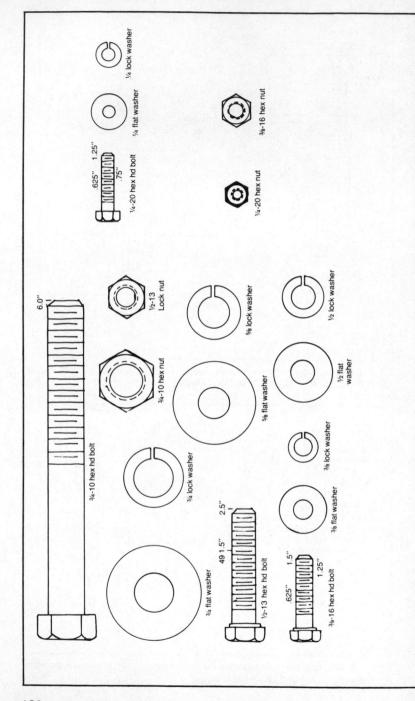

1/4 lock washer

1/4 flat washer

3/6-16 hex nut

625" 1.25"
.75"
1/4-20 hex hd bolt

1/4-20 hex nut

6.0"
3/4-10 hex hd bolt

1/2-13
Lock nut

3/4-10 hex nut

3/4 lock washer

3/4 flat washer

5/16 lock washer

5/8 flat washer

1/2 lock washer

1/2 flat washer

3/8 lock washer

3/8 flat washer

49 1.5"
2.5"
1/2-13 hex hd bolt

.625" 1.5"
1.25"
3/6-16 hex hd bolt

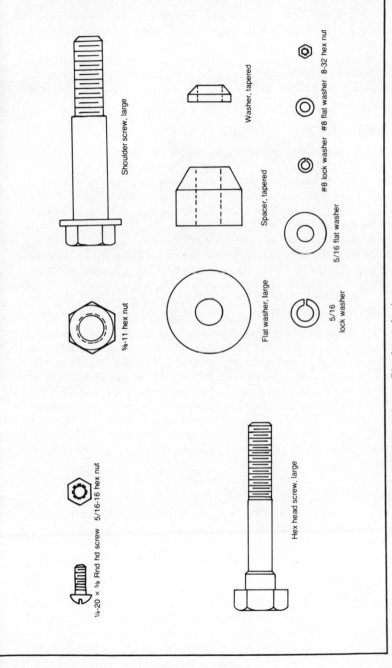

Fig. 7-16. Hardware for dish assembly. (Channel Master, Div. Avnet, Inc.)

hard. Cement dries from the top down, and what you may be feeling is the thin, upper layer.

Avoid mounting the dish near a hot radiating surface, such as the sun warmed side of a building or the black asphalt surface of a driveway, as the thermal energy will have an impact on in-band noise.

In some locations, any type of foundation must bear the seal of a professional engineer registered in the state in which the dish is to be erected. It is always preferable to check local regulations and building codes before starting any construction.

## MANUAL ADJUSTMENT OF THE DISH

Some dishes are intended for manual adjustment and so the dish needs to be turned to be able to receive signals from a number of satellites. Turning is done by a jack accompanied by a scale. Using the scale gives the operator assurance of repeatability—that is, being able to find a satellite that has been located previously. Since the scale and jack are outdoors they will need some form of weather protection.

At present, reception of all the North American and Canadian satellites requires the dish to swing about 90 degrees in azimuth. This does not mean 45 degrees on either side of true south. The amount of left/right movement will depend on location.

## INSTALLATION INSTRUCTIONS

If you plan to do your own dish installation, you will find a complete set of instructions accompanying this component. There is no universal set of instructions and so dish assembly will not be described here. However, there are certain guidelines you can follow with any dish installation.

One of the first steps is to take an inventory to make sure you have received everything. It is always helpful to have one or two assistants, but make sure they do not handle any components unnecessarily. You will find a collection of miscellaneous hardware included, and so it is advisable to become familiar with it. Figure 7-16 shows the more common types of hardware.

## COAXIAL CABLE

If you buy a complete TVRO system from one manufacturer you will be supplied with the coaxial cable for interconnecting components, or you will at least be advised as to the type of coaxial cable to use.

The 75-ohm cable for carrying the i-f signal between the downconverter mounted on the dish and the in-home satellite receiver will usually be RG 59U in most installations. Only high quality coax should be used, for this assures adequate shielding and prevents possible interference pickup. While RG 59U can be used for distances up to 300 feet between the receiver and downconverter, use RG 11U for longer distances. The wires used to deliver voltage to the downconverter should be gauge #20 for distances up to 300 feet, gauge #18 for 300 to 500 feet and gauge #16 for distances between 500 and 1,000 feet.

## CONNECTORS

A connector is any device for the easy connection and disconnection of coaxial cables or for connecting a cable to a component such as a satellite receiver. Connectors can be male or female. A male connector is a plug; a female a jack. However, there are many different kinds of plugs and jacks, and they must be designed to mate with each other. Since TVRO equipment is supplied with jacks, it is the user's responsibility to make sure the connecting cables are equipped with plugs that will fit into these jacks. While components are fitted with jacks, it is also possible to have coaxial cable fitted with either plugs or jacks.

Plugs and jacks are used not only with the TVRO system, but with associated components such as a videocassette recorder, a video disc system, and video cameras. Plugs are also used extensively for hi/fi systems that are sometimes used to handle the audio output of a stereo processor.

Figure 7-17 illustrates a pair of commonly used plugs. The one shown at the left is called a mini; the other is known as an RCA and is named for the manufacturer. These two plugs work in the same way, but aren't compatible, since they both require jacks of different sizes.

The advantage of using plugs and jacks is that it permits the immediate joining of one cable to another or a cable to a component, or as a means of connecting one component to another. There are two serious disadvantages: the first is the lack of compatibility and the other is that there are so many jacks and plugs, with some resembling each other closely, that connectors can be quite confusing. Further, a component used in a TVRO system could use two or more different types of connectors. Thus, a satellite receiver could use a type F connector for its input from the downconverter and a phono jack for its video and audio outputs. The drawing in Fig. 7-18 illustrates a number of different types of connectors. These include the F plug (A); the RCA plug (B); the mini plug (C); and the Motorola plug (D). Note that the RCA plug and the mini plug are different from those shown earlier in Fig. 7-17. This indicates that there are several varieties of mini and RCA plugs. While they may have the same names they are incompatible.

The F plug in Fig. 7-18 is extensively used in video components and on

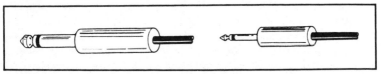

Fig. 7-17. Types of plugs. The one at the left is a mini; the other is an RCA plug. (Recoton Corp.)

183

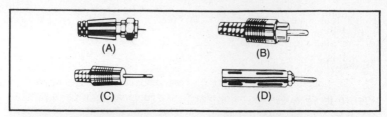

Fig. 7-18. Commonly used plugs. F plug, (A); RCA plug, (B); mini plug, (C); Motorola plug, (D). (Recoton Corp.)

coaxial cable. The RCA plug shown earlier in Fig. 7-17 is commonly used for audio work while the RCA plug in Fig. 7-18 is intended for both audio and video. The Motorola plug in Fig. 7-18 is for use in the wall plate from master antenna systems commonly found in high rise apartments.

A coaxial cable can be terminated in plugs, only, in jacks only, or in a plug at one end and a jack at the other, or it may be *raw* coaxial cable, with no connectors at either end. Further, if the coax has connectors, these plugs and jacks can be different. Figure 7-19 shows a variety of possibilities.

It is always advisable to avoid sharp bends in coaxial cable. If the jacks on the components to be connected face in the same direction the coaxial cable joining the two components will inevitably make a sharp bend, especially if the coaxial cable has a short length. This can be avoided by using a right-angle quick connect F plug as indicated in Fig. 7-19G. While this cable uses a pair of F plugs, note the physical differences between the two. This is another indication that plugs having the same name may not have identical physical characteristics.

A cable may be equipped with a pair of plugs at both ends as shown in Fig. 7-20H. This type of cable can be used when a pair of identical signal outputs of one component are to be connected to a pair of identical signal inputs of another component. Drawing I shows an F plug connected to an RCA plug and drawing J is an RCA plug wired to a 3.5 mm (35 cm) plug.

## COAXIAL CABLE LENGTHS

While it is generally desirable to have video components in the home positioned near each other, space availability may not always be enough to make this possible. Coaxial cable is often used for interconnecting components since such cable is less susceptible to noise and hum pickup. Coaxial cables equipped with jacks and plugs are available in different lengths such as 3, 6, 10, 15, and 25 feet (0.91, 1.82, 3.04, 4.57, and 7.62 meters).

When buying connecting cable, there are always several factors to consider. The first is to get a cable of suitable length. The cable should be long enough to permit making the connection but without sharp bending. Any passive device in the signal path, and that is exactly what coaxial cable is, introduces a certain amount of signal loss. For short lengths of cable carrying an adequate signal this is inconsequential.

184

Another factor is the type of cable selected. Coaxial cable is identified by RG numbers, with each cable having not only differing electrical characteristics, but different costs per unit length. For interconnecting components, it is advisable to follow the recommendations of the manufacturers of the TVRO equipment.

## HOW TO JOIN CONNECTORS AND COAX

It may not be easy or possible to get cable having the desired connectors or the available cable might have the right connectors but may not have the right length. It is easy to buy the right length of raw coaxial cable and the right connectors, but this produces the problem of joining the two.

One of the easiest connectors to use on coaxial cable is a screw-on type. As a start, examine the illustration in Fig. 7-20, which shows a

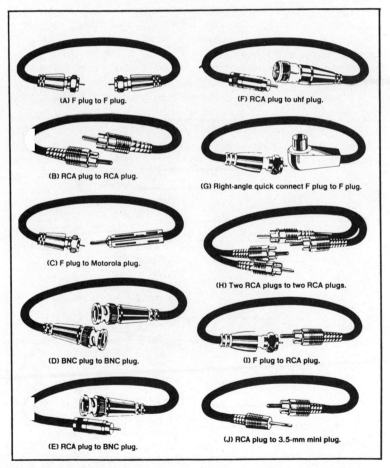

(A) F plug to F plug.

(B) RCA plug to RCA plug.

(C) F plug to Motorola plug.

(D) BNC plug to BNC plug.

(E) RCA plug to BNC plug.

(F) RCA plug to uhf plug.

(G) Right-angle quick connect F plug to F plug.

(H) Two RCA plugs to two RCA plugs.

(I) F plug to RCA plug.

(J) RCA plug to 3.5-mm mini plug.

Fig. 7-19. Plug combinations. (Recoton Corp.)

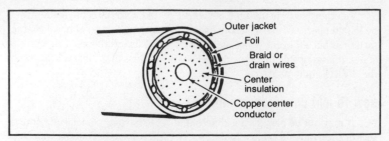

Fig. 7-20. Cross-section of unbalanced coaxial cable. (Winegard Co.)

cross-section of cable. It consists of an outer insulating jacket, a flexible rubber-like plastic material. Beneath this jacket you may find a layer of foil, although not all coaxial cable is so equipped. Beneath the foil, if the cable is so supplied, is a layer of flexible braid, sometimes referred to as a drain or drain wiring. The braid is one of the signal conductors and also works as an electrical shield. The braid surrounds a layer of insulating material, often flexible plastic, known as a dielectric. Finally, there is the center conductor, usually a solid wire rather than stranded wire. The braid is also called the cold lead; the center conductor the hot lead.

Figure 7-21 shows the steps to take to put a connector on coaxial cable.

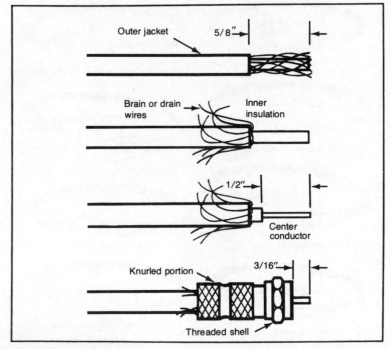

Fig. 7-21. Method of putting a connector on coaxial cable. (Winegard Co.)

186

Remove ⅝″ of the outer jacket and push the braid or drain wires back over the jacket. Remove ½″ of the inner insulation being careful not to nick the center conductor. Make sure the foil covering the inner insulation or the braid does not touch the center conductor. Now hold the connector by its knurled portion and twist it clockwise onto the end of the cable. The connector can be considered to be in place when the center conductor protrudes about ³/₁₆″ from the end of the cable and the inner insulation is flush at the back of the threaded shell.

## COAXIAL CABLE PRECAUTIONS

Because coaxial cable is passive and because its cost is so low in comparison with units such as a dish, LNA, or satellite receiver, it is often not regarded as important. Yet, the wrong cable, one inferior construction, or one with poorly joined connectors can often make the difference between a good TVRO system and one that is marginal.

Not all cable connections are of equal importance, but the one between the LNA and the downconverter requires the utmost attention. Between these two components the frequency is that of the downlink transmission and since this is measured in gigahertz, is highly critical. Ideally, this link should be as short as possible and as secure as possible. The wiring is often RG 8U or RG-213 cable using N connectors (Fig. 7-22).

Connectors are fastened to coax in a number of ways. The one shown earlier in Fig. 7-21 is a force-fit type. With some, a crimping tool is used to ensure a good connection between the connector and the braid. In some instances, an installation technician will simply use an ordinary pair of pliers, an undesirable practice. However, there are some pliers made for crimping. The best arrangement is one in which the braid and the connector are electrically joined by soldering.

There are several problems involving connectors. One of these is the poor fit made between the *hot* lead of the coax and the female fitting.

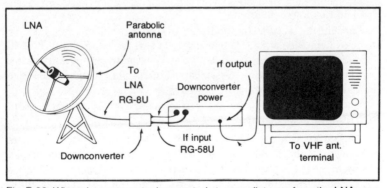

Fig. 7-22. When downconverter is mounted at some distance from the LNA, use RG-8U coaxial cable. The connecting line between the downconverter and the satellite receiver can be RG-58U coax. (Lowrance Electronics, Inc.)

187

Another problem is plug/jack combinations that have managed to work their way loose. Still another is the possibility of water getting into the connection. Special commercially available sealing compounds made specifically for this purpose are available. Coaxial cable from the outdoor downconverter to the in-home satellite receiver should be run through polyvinyl chloride (PVC) pipe. At the areas where the coaxial cable enters or leaves the pipe there should be a seal to prevent the entrance of water. The pipe can be buried in a trench about 18″ below ground and should extend from the dish pad to the point of entrance into the home.

Whenever moisture is present in any outdoor connector that carries dc voltage, carbon will develop. The only way to prevent this is to seal the connectors completely. Use an electrically inert compound such as Dow Corning #4 to prevent a water trap and subsequent accumulation of moisture.

Carbon deposits in the connectors can produce sparkles in the picture, either permanently or on an intermittent basis, a problem that is very difficult to trace.

## POWER CONSUMPTION

While a TVRO system consists of a number of components, these are all solid-state, and require very little energy. Some of the electrical energy needed is used for operating one or more motors, but the usage time is very small. The dish requires no electrical energy at all, nor does the scalar feed or the waveguide. The active components (the LNA, the downconverter, the satellite receiver, and the stereo processor) combined use less power than a typical television receiver having a 21″ screen.

## PREVENTING WATER DAMAGE

A downconverter may be weather resistant but not waterproof. It should be mounted in such a way that water does not run directly onto it. All cables should have drip loops to prevent water from flowing down cables and into the downconverter.

Sealing the rf connectors and terminal strip with silicone seal, RTV, or its equivalent is desirable, but do not try to seal the entire unit. This will cause water to be drawn inside. Water damage to the downconverter caused by neglect or improper installation can result in component damage not covered by the warranty.

## POSITIONING COMPONENTS

A satellite receiver can be used by itself, but to get greater enjoyment and control of the satellite signals, additional components are desirable. One of these is the stereo processor, described earlier and the other is a component for positioning the dish for reception from a selected satellite. One such component is a Memory Trak (described earlier in Fig. 6-9) that helps the dish *remember* its position for the specific location of each satellite. A readout panel indicates the satellite whose channels are being received.

If the three components are the same size, they can be stacked as shown in the photo in Fig. 7-23. Since all the controls of each of the components are near each other, finding a channel or selecting a sound program can be done quickly. The three components can be mounted near the TV set and an entertainment-center rack of the type used for hi/fi systems is desirable. For an installation of this kind, the TV receiver, if it is not a console type, can be mounted on top.

One other advantage of having the three components mounted as shown in the photo is that interconnections are short and easy to make.

## MULTIPLE SATELLITE TV SHARING

The use of a satellite TVRO setup isn't limited to a single family. A group of neighbors can share a single dish, LNA, and block downconverter. However, each home tied into the system would need to have a receiver capable of independent channel selection, with the understanding that selection of a particular satellite would need to be done by prior agreement. Using a block downconverter, the neighbors could each watch a different program, but all of these would need to come from a single satellite.

Signal sharing does result in a loss of signal strength, just as connecting two or more TV receivers to the same antenna produces a decrease in

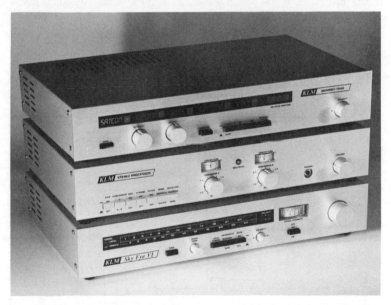

Fig. 7-23. Method of mounting satellite components. At the bottom the Sky Eye VI satellite receiver; center, the Stereo Processor and top, the Memory Trak. (KLM Electronics, Inc.)

signal strength. However, this can be overcome through the use of a line amplifier connected following the block downconverter.

Actually, the multi-use concept isn't new and is basically the same idea used by cable companies. They pick up the satellite signal via a dish and deliver that signal to their subscribers.

## SYSTEM VARIATIONS

The television receiver is often the center of a number of different video sources. The TVRO is just one of these, important because of the large number of programs it offers and the substantial possible improvement in picture quality. Other picture sources include earthbound television broadcasting, video discs, videocassette recorders, video games, and video cameras. The block diagram in Fig. 7-24 shows how these are related to the television receiver.

There are a large number of possible variations. These can be in the TVRO system, or can consist in using a component TV system instead of an integrated TV set. Thus, the component TV arrangement could comprise a separate TV tuner, a monitor, and a hi/fi system instead of an integrated TV receiver.

As far as the TVRO is concerned, it can be simple or sophisticated. Part of the variation depends on the dish, that is, on the type of mount that is used, on whether you want single channel downconversion or block downconversion, if you prefer a memory control of dish position, if you want a stereo processor, and if you are making plans for multiple TV viewing. You may also want to have your TVRO join forces with a high-fidelity sound system (Fig. 7-25).

## TVRO WITH MULTIPLE TV SETS

A single TVRO setup can be used to operate two or more television

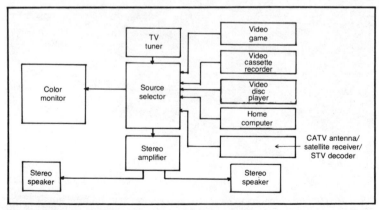

Fig. 7-24. One possible arrangement of an in-home entertainment system. (Zenith Radio Corp.)

190

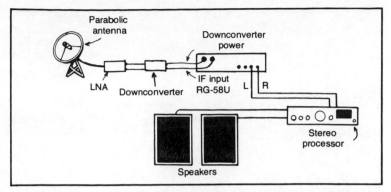

Fig. 7-25. The output of the Stereo Processor can be connected directly to a pair of high efficiency speakers or to the pre-amplifier input of a hi/fi system.

sets, as shown in the drawing in Fig. 7-26. These are connected following the remodulator in the satellite receiver, or a separate modulator if one is used. The hookup is very simple. The cable from the receiver goes to a two-way or four-way signal splitter and from the splitter to each of the TV receivers. More TV sets can be *daisy chained* with these splitters, as desired.

While this arrangement is easy, it does have a few limitations. The signal strength at the output of the receiver must be adequate. If it is marginal, the picture will be poor at each of the receivers. Each of the TV sets must have its channel selector set to either Channel 3 or Channel 4, but whichever of these is chosen, each TV set must be set to the same channel. Further, the same program will be watched on all the screens.

An alternative approach can be used as shown in Fig. 7-27 and Fig. 7-28. The component consists of a pair of independent modulators, both housed in the same cabinet.

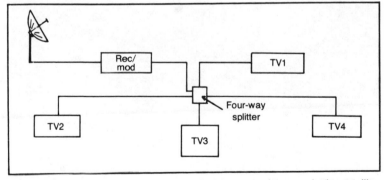

Fig. 7-26. Setup for operating four television sets from a single satellite receiver/modulator. For just two sets, use a two-way line splitter instead of the four-way unit shown here. RG-59 coaxial cable is generally used at the input and output of the splitter.

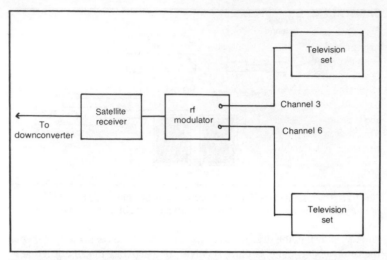

Fig. 7-27. Rf modulator shown in Fig. 7-28 can supply modulated signals on two different VHF channels to a pair of television receivers. The same satellite channel can be watched on both TV sets. The rf modulator is actually two modulators housed in one cabinet.

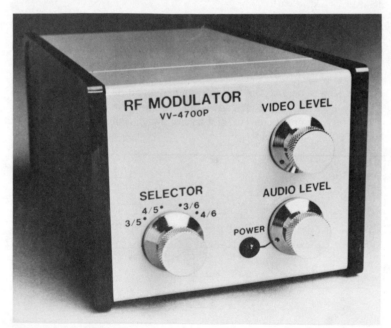

Fig. 7-28. Rf modulator. Unit contains two independent modulators, and can modulate baseband signals for Channels 3, 4, 5, and 6. It permits watching two different rf channels at the same time. However, the same program will be seen on both TV sets. (Showtime Video Ventures)

## THE DISH AND ITS ENVIRONMENT

One or more family members may object to a dish on the basis of appearance or location. Part of the problem is that the use of a dish can be a new experience and so may be rejected on that score.

From an appearance point of view, mesh dishes are somewhat more desirable for several reasons. They do not have the stark, often solid white appearance of fiberglass coated dishes. Since they are opaque, their solidity often makes them appear quite formidable. The mesh dish permits viewing right *through* the dish and so tends to blend in more with the environment. Further, mesh dishes are available in a number of colors, with a particular color selected for its blending ability. Figure 7-29 is a photo of a dish placed near the area occupied by a swimming pool. The dish is positioned in space that is not otherwise utilized. Bushes and plants can be placed behind or in front of the dish mount, provided those in front do not

Fig. 7-29. Dish blends in with swimming pool in limited backyard space. (North Star)

Fig. 7-30. The X-11, an 11-foot mesh dish in a garden environment. (KLM Electronics, Inc.)

interfere with the 'view' the dish has of the satellites. For home use a good location is at the rear of the home, not in front.

Even with space limitations, a dish can be made to blend in with the environment, as shown in Fig. 7-30.

## FIELD OF VIEW

The dish must have an unobstructed view of every satellite whose signals you want to receive (Fig. 7-31). This field of view depends on your latitude and longitude. Typical coordinates for satellite Satcom 3R and Comstar D2 are shown in Fig. 7-32 and Fig. 7-33. These are at opposite ends of the field of view and if they are *visible* with no obstruction between, all other satellites will be available.

You must also determine true north (Fig. 7-34) for your specific location. This is different from magnetic north found on a standard compass. Call your local airport, or your municipal building, or speak to the reference librarian in your local library for the compass correction for magnetic north, expressed in degrees east or west of true north. In some instances, the manufacturer of your dish may have this information available. If the compass correction figure is east subtract that number from the azimuth coordinate. If it is west, add it to the coordinate. Figure 7-35 shows these approximate values, but these change gradually so it is preferable to check your local sources.

## EXAMPLE

Let us assume that you are living in Eugene, Oregon and want to find F3, that is, Satcom 3R. To do this, you must have three pieces of information at hand: the latitude and longitude of Eugene, Oregon, and the compass correction. For this location, the latitude is 44 degrees; the longitude is 123 degrees and the compass correction is 20 degrees East.

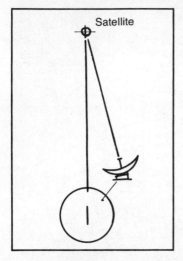

Fig. 7-31. Dish must have unobstructed view of satellite.

From the table of coordinates shown in Fig. 7-36, the azimuth for this location is 191 degrees and the elevation is 39 degrees. The correct compass heading for satellite F3 is found by subtracting the correction from the azimuth coordinate. Thus, $191 - 20 = 171$ degrees on a compass.

For this example, then, we can find F3 by aiming the dish at 171 degrees on the compass, with the dish pointed 39 degrees into the sky (Fig. 7-37). This satellite is a good satellite to seek as a starting point, as it has all 24 channels available. They are active and can be identified easily. You can now find Satcom F4 (Satcom 4) in the same manner.

Your local field of view is between these two satellites at the correct elevation. For any location, the satellite nearest true south will have the highest value of elevation (Fig. 7-38). Satellites to the extreme east or west will have a lower elevation. These calculations need be done only once for any area.

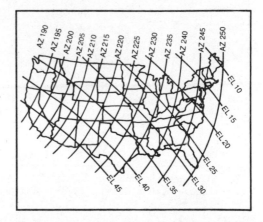

Fig. 7-32. Az/El coordi-
nates for Satcom F3(R),
131°W.

195

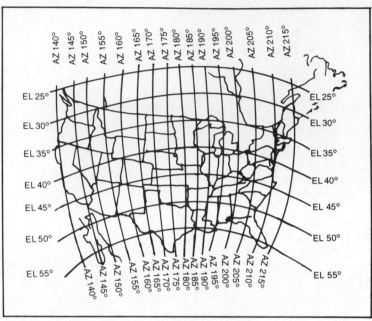

Fig. 7-33. Az/E1 coordinates for Comstar D2, 95°W.

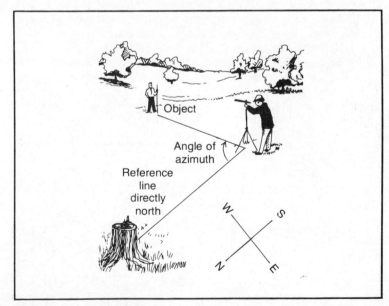

Fig. 7-34. To determine the angle of azimuth, a reference line running from true south to true north is established. All angles of azimuth are made with respect to this reference.

196

| Location | Correction | Location | Correction | Location | Correction |
|---|---|---|---|---|---|
| Alabama | 2E | Kentucky | 1E | North Dakota | 11E |
| Alaska | 26E | Louisiana | 6E | Ohio | 3W |
| Arizona | 14E | Maine | 20W | Oklahoma | 9E |
| Arkansas | 6E | Maryland | 8W | Oregon | 20E |
| California | 17E | Massachusetts | 15W | Pennsylvania | 8W |
| Colorado | 14E | Michigan | 3W | Rhode Island | 15W |
| Connecticut | 13W | Minnesota | 6E | South Carolina | 2W |
| Delaware | 10W | Mississippi | 5E | South Dakota | 11E |
| Washington D.C. | 8W | Missouri | 6E | Tennessee | 1E |
| Florida | 2E | Montana | 18E | Texas | 10E |
| Georgia | 0 | Nebraska | 11E | Utah | 15E |
| Hawaii | 11E | Nevada | 17E | Vermont | 15W |
| Idaho | 19E | New Hampshire | 16W | Virginia | 6W |
| Illinois | 2E | New Jersey | 11W | Washington | 22E |
| Indiana | 0 | New Mexico | 13E | West Virginia | 5W |
| Iowa | 6E | New York | 10W | Wisconsin | 2E |
| Kansas | 9E | North Carolina | 5W | Wyoming | 13E |
| Alberta | 22E | Manitoba | 10E | Saskatchewan | 17E |
| British Columbia | 23E | Ontario | 8W | Quebec | 17W |

Fig. 7-35. Magnetic compass corrections for North America. (Total Television, Inc.)

| State | City | Declination Angle | Elevation | Azimuth |
|-------|------|------------------|-----------|---------|
| AL. | Birmingham | 5.4 | 29 | 240 |
| ARIZ. | Phoenix | 5.4 | 45½ | 212 |
| ARK. | Little Rock | 5.7 | 30½ | 235 |
| CA. | S.F. | 6.1 | 45 | 193 |
| CA. | L.A. | 5.5 | 48 | 202 |
| CO. | Denver | 6.3 | 36½ | 216 |
| CONN. | Hartford | 6.6 | 15 | 247 |
| RI. | Providence | 6.6 | 13 | 249 |
| DEL. | Wilmington | 6.3 | 17 | 247 |
| MD. | Baltimore | 6.2 | 19 | 245 |
| FL. | Jacksonville | 5.0 | 27 | 246 |
| FL. | Miami | 4.4 | 27 | 250 |
| GA. | Atlanta | 5.5 | 27 | 242 |
| IDA. | Boise | 6.8 | 37 | 201 |
| ILL. | Chicago | 6.6 | 25 | 234 |
| IN. | Indianapolis | 6.3 | 25 | 237 |
| IOWA | Des M | 6.6 | 28 | 229 |
| KANSAS | Wichita | 6.1 | 33½ | 228 |
| KY. | Louisville | 6.1 | 26 | 238 |
| LA. | Baton Rouge | 5.0 | 34 | 239 |
| MAINE | Bangor | 6.9 | 11 | 249 |
| MASS. | Boston | 6.6 | 13 | 249 |
| MICH. | Grand Rapids | 6.7 | 23 | 236 |
| MICH. | Marquette | 7.1 | 21 | 233 |
| MINN. | St. Paul | 6.9 | 26 | 228 |
| MISS. | Jackson | 5.3 | 32½ | 239 |
| MISSOURI | Kansas City | 6.2 | 31½ | 229 |
| | St. Louis | 6.2 | 27½ | 235 |
| MONT. | G. Falls | 7.2 | 31½ | 206 |
| NEB. | Gnd. Island | 6.5 | 32 | 225 |
| NEV. | Las Vegas | 5.8 | 45 | 206 |
| NH. | Laconia | 6.7 | 13 | 248½ |
| NJ. | Trenton | 6.3 | 16 | 245 |
| NM. | Alb | 5.7 | 42 | 219 |
| NY. | Syracuse | 6.7 | 16½ | 244½ |
| NC. | Durham | 5.8 | 22 | 245 |
| ND. | Bismark | 7.1 | 29 | 218 |
| OHIO | Columbus | 6.3 | 23 | 240 |
| OK. | Ok. City | 5.7 | 36½ | 228½ |
| OR. | Eugene | 6.8 | 39 | 191 |
| PA. | Harrisburg | 6.3 | 18½ | 245 |
| SC. | Columbia | 5.5 | 24 | 245 |
| SD. | Rapid City | 6.8 | 32 | 217 |
| TENN. | Memphis | 5.7 | 31 | 237 |
| | Knoxville | 5.8 | 25½ | 241 |
| TEXAS | Abilene | 5.3 | 40 | 229 |
| UTAH | Salt Lake City | 6.5 | 39 | 208 |
| VERMONT | Montpelier | 6.8 | 13 | 247 |
| VA. | Charlottville | 6.1 | 21 | 244 |
| WASH. | Spokane | 7.2 | 33 | 198½ |
| W. VA. | Charleston | 6.2 | 23 | 242 |
| WIS | Eau Claire | 6.9 | 25 | 230 |
| WYO. | Casper | 6.7 | 34 | 214 |

Fig. 7-36. Declination angle, elevation, and azimuth for Satcom 3R(F3). (Total Television, Inc.)

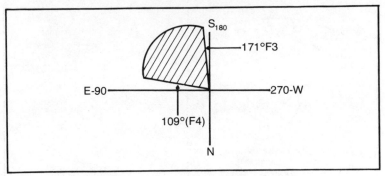

Fig. 7-37. Field of view for Eugene, Oregon. (Total Television, Inc.)

## ANGLE OF DECLINATION

All satellites pursue a circular orbit, but this orbit is with reference to the equator. If you had a dish positioned on the equator, you could pick up all the satellites by rotating the dish in a circular manner. However, as the installed location gets further from the equator the satellites appear in an elliptical orbit (Fig. 7-39) and simply rotating the dish circularly will not find them all. Your dish will probably have a mechanical arrangement to permit it to swing in an elliptical track.

## SATELLITE ECLIPSE

Each spring and fall, a satellite receiving system will experience a phenomenon known as satellite eclipse. This occurs when the sun positions itself directly behind a satellite and an eclipse of the sun occurs. When this happens you can see the shadow of the LNA in the exact center of your dish. These periods of *eclipsing* occur for five to seven days and will affect reception in varying degrees for about one hour each day.

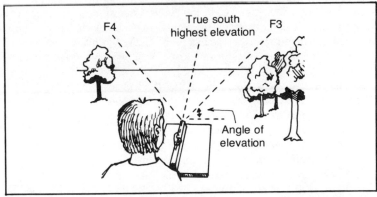

Fig. 7-38. True south represents the highest elevation. (Total Television, Inc.)

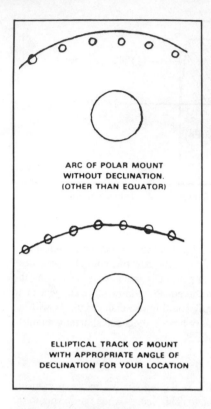

ARC OF POLAR MOUNT
WITHOUT DECLINATION.
(OTHER THAN EQUATOR)

ELLIPTICAL TRACK OF MOUNT
WITH APPROPRIATE ANGLE OF
DECLINATION FOR YOUR LOCATION

Fig. 7-39. To a dish located north of the equator, the satellites appear to be in an elliptical orbit. (Total Television, Inc.)

## PERSONAL INSTALLATION

Installing a TVRO system isn't difficult, since manufacturers do supply completely detailed setup information. The pad used as my dish support measures 4' × 4' and is 2' deep. The cement fill doesn't rest on the ground, but is on a buried wire mesh having the full dimensions of the opening. The mesh rests on crushed stone. In the center, almost 3' down, there is a substantial cement block and on top of this is a cinder block. The mounting pipe was forced through the opening of the cinder block and then, using a two-foot long bubble glass, was moved until the pipe was absolutely vertical.

The cement was mixed using a rented, electrically operated cement mixer. After the cement was in position, a number of checks with the bubble glass were taken to make certain the mounting pipe had not moved and was still at a right angle with the earth.

The cement was allowed to harden for five days and was sprinkled lightly with water every day as part of the cement curing process. No one was allowed to walk on the cement during this time.

More than a half-century ago, toys for children often included an Erector Set, generally for boys only. Times have changed and today's

Fig. 7-40. Sliding the pedestal over the mounting pipe. (Daniel and Susan Clifford)

Fig. 7-41. Fastening the elevation arm to the pedestal.

generation of boys and girls are much more sophisticated and are able to take the assembly of a satellite dish in stride.

Figure 7-40 shows the initial step in the dish setup. The pedestal has been slipped over the support pipe and was then locked into place. The elevation arm (Fig. 7-41) was easily mounted and fastened. The next photo, Fig. 7-42, shows the tangential drive ring being put into position.

The lower hub plate is now in place and the two young helpers are busy

Fig. 7-42. The tangential drive ring is put into position and fastened.

Fig. 7-43. The tripod legs are securely attached to the upper hub plate.

at work on the upper hub plate (Fig. 7-43). To make the work easier, the assembly is being done on a folding table.

Figure 7-44 shows the ribs of the dish being fastened into position. No measuring is required with pre-drilled holes making rib location easy (Fig. 7-45). Mounting the ribs becomes a group project (Fig. 7-46).

Finally, the dish begins to take shape with mesh panels sliding easily

Fig. 7-44. Putting the ribs in place. (From left to right: Martin Clifford, Susan and Daniel Clifford, and Brian Reck)

into place (Fig. 7-47). The petals are held securely with perimeter channels (Fig. 7-48).

Figure 7-49 shows the drive motor being put into position. The project nears completion as the wiring is fastened to a terminal block.

Fig. 7-45. The rib assembly is easily supported by a small table.

204

Fig. 7-46. Final steps in the rib assembly.

Figure 7-50 shows the working crew and their supervisors in front of the finished dish, and just in time for lunch. The wiring consisted of coaxial cable and a multiconductor cable for supplying operating voltages for the LNA, the downconverter, and the motor. The cables were brought through

Fig. 7-47. With the rib assembly completed, the mesh petals slide easily into place. (Courtland Newton, III at the left, and Brian Reck)

Fig. 7-48. Under the watchful eye of Gary Gordon, Production Coordinator for KLM, Courtland fastens a perimeter channel.

Fig. 7-49. The drive motor is put into position.

Fig. 7-50. The working crew after completion of dish assembly. From left to right: Gary Gordon of KLM, Daniel Clifford, Brian Reck, Martin Clifford, Courtland Newton, III, Adrienne Clifford, Susan Clifford and Courtland G. Newton, Jr. of the Daniel Roher Co.

10-foot sections of PVC pipe having an inside diameter of ¾″ and buried in a one foot trench. At the house, the wires were brought up the outside wall, onto the roof. A hole was drilled in the roof and fitted with a right-angle section of pipe ending in a flange. Inside the house, the wires were brought through the attic, and were then snaked down a wall, exiting at a baseboard near the television receiver. Reception is excellent and was so right from the start.

# Chapter 8

# Direct Broadcast Satellite

Direct broadcast satellite, also known as DBS, is on its way. Presently, if TV viewers want satellite reception they have two choices. One is to make use of cable TV service, assuming such service is available in their area, and the other is to set up their own TVRO system.

Cable TV always involves some payment on the part of the TV watcher for the installation of an rf signal converter/channel selector, and also for certain selected programs. There may also be an additional charge for a decoder in the event some of the program material is scrambled.

Cable TV works somewhat like a middleman, picking up satellite signals and routing them to home viewers via coaxial cable. Cable TV does more than that though, for cable companies amplify the video signals they receive, something for which there is no alternative, since they must supply thousands of customers. They also remove glitches and noise voltages. As a result, the home viewer not only gets more program diversity since satellites can supply many more programs than ordinary broadcast TV, but better video. Cable TV also eliminates the need for a home-type outdoor TV antenna, plus rotator, plus radio-frequency amplifier (or pre-amplifier) required for fringe area reception. With the arrival of DBS, cable companies will get competition other than that supplied by free broadcast TV.

## THE K BAND

DBS will be a satellite service, but the operating frequencies will be much higher than those used in the C band, and will be in the 12 GHz region. This is known as the Ku band, also simply referred to as the K band.

DBS is expected to have on orbital slot per time zone although there is some expectation that each time zone will have two orbital slots. As a consequence, although each DBS satellite will handle relatively few programs, possibly 3 or 4, it is expected that the total program capacity in the 12 GHz band will be approximately a minimum of 30 and a maximum of 90 channels for each time zone. Space allocation in the 12 GHz band for downlink transmissions is 500 MHz.

## DBS TRANSPONDER POWER

Instead of using small power transponders, such as the 5-watt types in the C band, some suggestions have been made that power will be in the order of 160 watts. While emphasis has been placed on the higher power to be used by DBS transponders, the actual amount of power hasn't been disclosed. The RCA DBS satellite is expected to use a low 40 watts, but even this is still eight times as much power used by transponders in the C band.

## BROADCAST VS SATELLITE TRANSMISSION MODE VS DBS MODE

All terrestrial TV transmissions are considered to be in the broadcast mode, that is, these video transmissions are not directed to any individual, group of individuals, any company, or group of companies. The video signals are transmitted, and anyone in any section of the geographical area covered by the TV station (or outside it, for that matter) is free to use a TV set to pick up such broadcasts. While video signals from TV satellites are also broadcast in the sense that they are transmitted they are intended for pickup by specific cable companies. The concept is that these signals are directed or targeted, even though there are thousands of cable companies scattered throughout the U.S.

DBS transmissions, however, even though made via a satellite, will be regarded from the start as TV broadcasts. This does not mean such signals cannot be scrambled. The act of scrambling does not determine whether a video signal is in the broadcast mode or not.

## TRANSPONDER OPERATING POWER

Transponders for K band satellites will get their operating power from the same source that supplies power for C band transponders. This will be from solar panels extending from the body of the satellite, furnishing electricity to storage batteries. However, because of the higher operating power requirements there will be fewer transponders, variously estimated from three to six, and because of this limitation, the number of available programs will also be limited.

## DISH SIZE

The use of higher power by the DBS transponders means the dishes for signal reception will be substantially smaller than those used presently on the C band and will have apertures of 2 to 4 feet. This means the gain of such dishes will be less but this is compensated by the use of higher transponder power.

A smaller dish has a number of advantages. It will be easier to manufacture a dish that will be truly parabolic and will be able to hold its shape. The smaller size means greater ease in shipping and less stress on the dish. Further, the smaller dishes will not need to tolerate such a high wind load factor and the smaller size and weight will permit the dishes to be mounted on a rooftop, or possibly on a window sill.

Since there will be far fewer satellites on the K band, dishes will probably be fixed position, not scanning types. However, the problem of locating a satellite will remain. For those who expect window sill installation, not all windows will be suitable, for those facing the *wrong* side will be unable to make use of dishes. While there may be less geothermal noise, heat reflected by building walls may increase the noise level. If DBS dishes are to be mounted on a roof, the dish, even though small by C band standards, will need to be secured. Further, the roof may need to be reinforced, not only to tolerate the weight of the dish, but wind force as well. When DBS finally does make its appearance, local governments may pass ordinances governing the installation of roof-top dishes.

The use of the K band is authorized by the Federal Communications Commission. However, the first DBS satellite is not expected to become operational until 1986.

## QUASI DBS

In the meantime, some companies are leasing space on satellites now under construction. These will be modified to use higher transmitting power and so the expected dish size will be reduced to 4 feet.

Unlike programs received in the C band, satellite signals in the DBS K band are expected to be pay TV or advertising oriented. Basically, DBS is intended for non-urban areas not now supplied with TV.

# Appendix
# System Design

This section is supplied for those who may be interested in TVRO system design. The material was extracted from a publication entitled Television Receive-Only (TVRO) Planner, produced by M/A-Com Video Satellite, Inc. and is reproduced here with their permission.

## INTRODUCTION

An acceptable video signal-to-noise ratio is the prime design criteria in designing an earth station. The definition of "acceptable" depends on the user and is a function of the TVRO and distribution losses after the satellite receiver. In a home application, the video signal-to-noise is virtually that of the earth station. In a cable system, the worst case video signal-to-noise ratio is that of the earth station reduced by miscellaneous losses between the earth station and the furthest subscriber.

The purpose of this section is to walk the reader through a downlink analysis. Useful approximations are mentioned where appropriate but are not used in this analysis as they reduce the accuracy of the calculations in which they are used and detract from the object of this exercise, which is to make accurate calculations.

At first, these calculations may seem complex and

211

confusing. They are not. Where possible, equations
are broken down into simpler expressions which are
more readily handled.

Satellite-transmitted signals are frequency modulated
(FM). Picture and sound quality are determined by a
quantity called the carrier-to-noise ratio or C/N. A C/N
ratio of 10 dB where carrier-to-noise is defined as:

$$C/N = \frac{P_{carrier}}{P_{noise}}$$

Where: $P_{noise}$ = noise power
$P_{carrier}$ = carrier power

is considered to be *FM threshold*. Above threshold
there is a 1 dB increase in video signal-to-noise ($V_{S/N}$)
for every 1 dB increase in carrier-to-noise (C/N). The
C/N ratio above threshold then, defines the $V_{S/N}$.

Standards for acceptable $V_{S/N}$ ratio vary. The Electronic
Industries Association (EIA) in *Electrical Performance
Standards for Television Relay Facilities (EIA RS 250B)*
sets the following $V_{S/N}$ standards for microwave
systems:

1. short haul – 53 dB
2. medium haul – 48 dB
3. **satellite – 50 dB**
4. long haul – 44 dB
5. end-to-end – 43 dB

## CALCULATING SIGNAL-TO-NOISE RATIO:

TO CALCULATE S/N:

1. Calculate $V_{S/N}$ as a function of C/N
2. Calculate $A_{S/N}$ as a function of C/N
3. Calculate C/N
4. Calculate S/N

## 1. Calculate $V_{S/N}$ as a Function of C/N

The video signal-to-noise ($V_{S/N}$) is expressed as the ratio of the luminance signal (in V P-P) to the noise signal (in V RMS). The luminance
portion of a 1V P-P video signal is 714 mV (100 IRE units) P-P.

212

The relationship between $V_{S/N}$ ratio and C/N ratio when the C/N ratio is above threshold is:

$$V_{S/N} = C/N \text{ (dB)} + 10 \log_{10} \frac{BW_{RF}}{2BW_m} + 9 \text{ dB}$$

$$+ 10 \log_{10} 3 \left(\frac{D}{BW_m}\right)^2$$

$$+ \text{ emphasis improvement}$$

Where: C/N is the carrier-to-noise ratio in dB.

$10 \log_{10} \dfrac{BW_{RF}}{2BW_m}$ = the AM and FM bandwidth

correction factor.

9 dB = the conversion from $\dfrac{RMS}{RMS}$ to $\dfrac{P-P}{RMS}$ S/N ratio.

$10 \log_{10} 3 \left(\dfrac{D}{BW_m}\right)^2$ is the FM improvement factor.

$BW_{RF}$ is the system noise bandwidth, or satellite receiver half-power IF bandwidth.

D is peak-to-peak deviation of FM carrier.

$BW_m$ is maximum frequency of baseband modulation signal.

emphasis improvement = 12.8 dB.

*EXAMPLE:* Calculate the $V_{S/N}$ as a function of a given satellite receiver and a variable C/N ratio. First, note that for a downlink signal D = 10.75 MHz peak deviation of FM carrier. $BW_m$ = 4.2 MHz bandwidth of baseband modulation signal and that $BW_{RF}$ is a function of the satellite receiver. In this example, we will use a typical number $BW_{RF}$ = 27 MHz = $2.7 \times 10^7$ Hz which is a satellite receiver's 3 dB IF bandwidth, and is roughly equivalent to its noise bandwidth. For a system using a 27 MHz bandwidth and the modulation parameter of this example C/N (dB) = desired $V_{S/N}$ (in dB) − 39.8. This equation can be used for other satellite receivers by using the appropriate value of $BW_{RF}$.

$$V_{S/N} = C/N \text{ (dB)} + 10 \log_{10} \frac{BW_{RF}}{2BW_m} + 9 \text{ dB} + 10 \log_{10} 3 \left(\frac{D}{BW_m}\right)^2 + 12.8$$

$$= C/N \text{ (dB)} + 10 \log_{10} \frac{27 \times 10^6}{2(4.2 \times 10^6)} + 9 \text{ dB} + 10 \log_{10} 3$$

$$\frac{10.75 \times 10^6}{4.2 \times 10^6} + 12.8$$

$$= C/N \text{ (dB)} + 5.07 + 9 + 12.93 + 12.8$$

$$= C/N \text{ (dB)} + 39.8$$

## 2. Calculate $A_{S/N}$ as a Function of C/N

The expression for audio S/N ratio (again above threshold) is:

$$A_{S/N} \text{ (dB)} = C/N + 10 \log_{10} \frac{3}{4} + 10 \log_{10} \frac{X^2}{F_{sc}^2}$$

$$+ 10 \log_{10} \frac{\Delta F_{sc}^2}{F_A} + 10 \log_{10} \frac{BW_{RF}}{F_A} + E_A$$

Where: C/N = the carrier-to-noise ratio in dB.

$X$ = deviation of carrier by subcarrier.

= 2 MHz.

$F_{sc}$ = peak deviation of subcarrier in Hz.

= 75 kHz.

= $7.5 \times 10^4$ Hz.

$F_A$ = highest audio baseband frequency in Hz.

= 15 kHz.

= $1.5 \times 10^4$ Hz.

$F_{sc}$ = subcarrier frequency in Hz. (e.g., $6.2 \times 10^6$ Hz; $6.8 \times 10^6$, etc.).

= $6.8 \times 10^6$ Hz (in this example).

$E_A$ = audio emphasis improvement.

$$= 13.2 \text{ dB (per CCIR Vol. XII, Rep. 637).}$$

$$BW_{RF} = \text{receiver noise bandwidth.}$$

$$= 27 \text{ MHz.}$$

$$= 2.7 \times 10^7 \text{ Hz (in this example).}$$

Rearranging the terms in this equation yields:

$$A_{SN} \text{ (dB)} = C/N + 10 \log_{10} \frac{3 X^2 \Delta F_{sc}^2}{4F_{sc}^2} \frac{BW_{RF}}{F_A^2} + E_A$$

Substituting values into this expression yields

$$A_{SN} \text{ (dB)} = C/N + 10 \log_{10}$$
$$\left[ \frac{3 (2 \times 10^6)^2 (7.5 \times 10^4)^2 (2.7 \times 10^7)}{4 (6.8 \times 10^6)^2 (1.5 \times 10^4)^3} \right] + 13.2$$

$$= C/N + 10 \log_{10} (2.919 \times 10^3) + 13.2$$

$$= C/N + 34.65 + 13.2$$

$$= C/N + 47.85.$$

Using this result, we can derive

$$C/N = \text{desired } A_{S/N} \text{ (dB)} - 47.85 \text{ for the modulation}$$
parameters given in the example.

We can now quickly calculate, for a given set of parameters which are all constants from the user's point of view, both $V_{S/N}$ and $A_{S/N}$ using the approximations:

$$V_{S/N} \text{ (dB) } C/N + 39.8$$

and

$$A_{S/N} \text{ (dB) } C/N + 47.85.$$

### 3. Calculate C/N

Consider the satellite-to-earth downlink as a very long microwave path (which indeed it is). The downlink carrier-to-noise (C/N) in dB can be calculated using the expression:

$$C/N = EIRP - S + G/T - 10 \log_{10} B + 228.6.$$

Where: EIRP = effective isotropic radiated power in dBw.

S = free space path loss between satellite and earth station in dB.

G/T = a quality factor characterizing the antenna/system noise temperature combination in dB/°K.

B = receiver noise bandwidth in Hz.

228.6 = Boltzman's Constant, a physical factor independent of path.

In the above expression, the parameters EIRP and S are functions of earth station locations. Both these and Boltzman's Constant are beyond the user's control. The bandwidth of the receiver, $BW_{RF}$ is a function of the signal which is to be received (which, in this case, is video), is determined by the satellite receiver, and must be wide enough to maintain picture quality (i.e., 25 to 36 MHz depending on application). The only remaining parameter is the earth station quality factor, G/T, the importance of which cannot be overemphasized, as it is the only factor over which the user has control.

To properly use the above expression to calculate downlink C/N, the system designer must determine EIRP, S, G/T, and $BW_{RF}$.

Calculations for S and G/T are presented below. Values of EIRP and $BW_{RF}$ may be determined from published footprint maps and manufacturer's specifications respectively.

## A. Calculate S

The length of the straight-line path between the satellite and the earth station is called the *SLANT RANGE*. As the satellite and the earth station are stationary with respect to each other at all times, this range is constant and has been calculated using geometry and the latitude and longitude of the satellite and the earth station. These calculations have been combined with the formula for free space path loss (in dB) for a given distance and frequency to yield the resulting expression:

$$S = 185.05 + 10 \log_{10} [1 - 0.295 \cos (TVROLAT)$$

$$\cos (|SATLONG - TVROLONG|)] + 20 \log_{10} F_{GHz}$$

Where: S = free space path loss (dB).

TVROLAT = latitude of TVRO.

SATLONG = longitude of satellite.

TVROLONG = longitude of TVRO.

|SATLONG − TVROLONG| = absolute value of difference between satellite longitude and TVRO longitude.

$F_{GHz}$ = frequency in GHz.

*EXAMPLE:* Determine the free space path loss from SATCOM I to an earth station at 42.5° west latitude and 71.15° north longitude. The longitude of SATCOM I is 135° west.

This problem is a case of plugging the correct values into the equation for free space path loss:

$$S = 185.05 + 10 \log_{10} [1 - 0295 \cos (TVROLAT)]$$

$$\cos (|TVROLAT - SATLAT|)] + 20 \log_{10} F_{GHz}$$

$$= 185.05 + 10 \log_{10} [1 - 0.295 \cos (42.5)$$
$$\cos (|71.15 - 135|)] + 20 \log_{10} (3.95)$$

$$= 185.05 + 10 \log_{10} [1 - 0.295 (0.73721)(0.440721)]$$
$$+ 20(0.596)$$

$$= 185.05 + 10 \log_{10} (1 - 0.0958) + 11.93$$

$$= 185.05 + 0.43 + 11.9$$

$$= 196.5 \, dB$$

Note: values of S typically lie between 195 and 197 dB. For quick calculations, the approximation $S \approx 196$ dB is adequate.

### B. Calculate G/T

G/T is a quality factor characterizing the antenna gain/system noise temperature in dB/°K. G/T *is the only variable* in the satellite link equation over which the system designer has practical control.

When designing a system, the system designer must trade off antenna gain and LNA noise temperature in the G/T calculations until he has achieved the desired G/T and hence C/N and $V_{S/N}$. All other parameters are essentially fixed.

217

The expression for G/T is:

$$G/T \approx G_a - 10 \log_{10} T_s$$

Where: $G_a$ = antenna gain in dB.

$T_s$ = system noise temperature in °K.

## I. Determine Antenna Gain

Antenna gain is determined from manufacturer's published data. However, calculating the gain of a 55% efficient parabolic antenna, the following approximation is useful.

$$G = 20 \log_{10} D + 20 \log_{10} F + 7.5$$

Where: D = parabola or parabolic antenna diameter in feet.

F = frequency in GHz.

Note: 55% efficiency is considered to be typical or slightly conservative.

## II. Calculate System Noise Temperature

The system noise temperature, $T_s$, contains contributions from the antenna, the waveguide or antenna feed, the LNA, the LNA-to-satellite receiver feedline, and the satellite receiver itself. We will now examine the expression for $T_s$ and how each element of the TVRO effects it.

$$T_s = T_{ant} + (L_1 - 1)T_0 + L_1 T_{LNA} + \frac{L_1(L_2 - 1)T_o}{G_{LNA}} + \frac{L_1 L_2 T_{RX}}{G_{LNA}}$$

Where: $T_{ant}$ = antenna noise temperature – a function of antenna size and look angle. For a typical 4 to 5 meter dish, $T_a \approx 88 \, El^{-0.39}$.
Where El is look angle above the horizon.

$L_1$ = contribution from losses between antenna and LNA.

$$= antilog_{10} \left[ \frac{loss \, (dB)}{10} \right]$$

$T_0$ = ambient temperature in °K.

218

$$= °C + 273.3$$

$T_{LNA}$ = LNA noise temperature in °K.

$L_2$ = contribution from feedline losses and power divider between LNA and satellite receiver.

$$= antilog_{10}\left[\frac{loss\ (dB)}{10}\right]$$

$T_{RX}$ = satellite receiver noise temperature in °K.

$G_{LNA}$ = LNA gain factor.

$$= antilog_{10}\left[\frac{gain\ (dB)}{10}\right]$$

Let's look at the equation again and examine the sources of the noise contributions:

$$T_s = T_{ant} + (L_1 - 1)T_0 + L_1 T_{LNA} + \frac{L_1(L_2 - 1)T_0}{G_{LNA}} + \frac{L_1 L_2 T_{RX}}{G_{LNA}}$$

$T_s$ = antenna + ambient temperature + LNA + feedline + receiver

Antenna noise temperature, $T_{ant}$, contains contributions from the antenna itself, thermal radiation from the earth, and sky noise pickup. The largest contribution is by thermal radiation from the earth; hence, the lower the antenna's elevation, the larger the antenna's noise temperature. In our example, we will use a 5 meter antenna with 44.4 dB gain and a $T_{ant}$ of 26° at 22° elevation.

Worst case ambient temperatures in the United States and Canada occur in the summer. A hot summer day in Maine might be 95°F (308°K), in Florida about 110°F (316°K).

The coefficient of $T_0$, $L_1 - 1$ is due to losses between the antenna and LNA which are typically on the order of 0.1 dB. The $L_1$ coefficient thus becomes

$$L_1 = antilog_{10}\frac{0.1}{10}$$

$$= 1.023$$

and

$$L_1 - 1 = (1.023 - 1)$$

$$= 0.023$$

The contribution to system noise temperature due to ambient temperature becomes:

contribution $= 0.023\ T_0$

The contribution to system noise temperature from the LNA is:

contribution $= L_1 T_{LNA}$

$$= 1.023\ T_{LNA}$$

Note that, for an LNA with a noise temperature of 120°K, this contribution becomes

contribution $= L_1 T_{LNA}$

$$= (1.023)(120)$$

$$= 122.76°K$$

Both feedline loss ($L_2$) and LNA gain figure in the term

$$\frac{L_1(L_2 - 1)T_0}{G_{LNA}}$$

The LNA gain term is

$$G_{LNA} = antilog_{10}\left[\frac{LNA\ gain\ (dB)}{10}\right]$$

and for a typical LNA gain of 50 dB becomes

$$G_{LNA} = antilog_{10}\left[\frac{50}{10}\right]$$

$$= 100,000.$$

For an LNA gain of 60 dB, $G_{LNA} = 1,000,000$.

The feedline loss then has to be calculated for the length and type of feedline used. First, the loss of the feedline run is calculated:

220

feedline loss = (loss per 100 Ft.)(cable length in 100's of Ft.)

Note that for cable lengths less than 100 feet, the feedline length is expressed as a decimal value less than 1, that is, 50 feet is 0.5 x 100 feet and appears as 0.5 in the above expression.

For 200 feet of 1/2 inch foam cable with a loss of approximately 8.5 dB/100 Ft.

feedline loss = 8.5 dB/100 ft. x 200 ft.

$$= 17 \text{ dB}$$

Calculating $L_2$ from feedline loss (no power divider in system)

$$L_2 = \text{antilog}_{10} \left[ \frac{\text{loss (dB)}}{10} \right]$$

$$= \text{antilog}_{10} \left[ \frac{17}{10} \right]$$

$$= 50.118$$

The feedline contribution to system noise temperature becomes

$$\text{feedline contribution} = \frac{L_1(L_2 - 1)T_0}{G_{LNA}}$$

$$= \frac{L_1(L_2 - 1)T_0}{G_{LNA}}$$

$$= \frac{1.023 \,(50.118 - 1)\, 315}{100,000}$$

$$= 0.158°K$$

For the same length of 7/8" foam feedline (loss = 5.2 dB/100 Ft.), the feedline contribution would be reduced to

$$\text{feedline contribution} = \frac{L_1(L_2 - 1)T_0}{G_{LNA}}$$

$$= \frac{1.023 \,(10.96 - 1)315}{100,000}$$

$$= 0.032°K$$

What does this mean? Feedline losses under 20 dB are "washed" out by the 50 dB (100,000) LNA gain. You can use the cheaper cable. We will again show this in the receiver contribution term.

The last term of the system noise temperature contains the contribution from the satellite receiver noise temperature:

receiver contribution $= L_1 L_2 T_{RX}$

For a satellite receiver noise temperature of 14,000°K (approximately 17 dB noise figure) and other parameters as determined above

$$\text{receiver contribution} = \frac{L_1 L_2 T_{RX}}{G_{LNA}}$$

$$= \frac{(1.023)(50.118)(14,000)}{100,000}$$

$$= 7.17°K$$

If the 1/2" foam feedline is used instead of the 7/8", the receiver contribution becomes

$$\text{receiver contribution} = \frac{L_1 L_2 T_{RX}}{G_{LNA}}$$

$$= \frac{(1.023)(10.96)(14,000)}{100,000}$$

$$= 1.56°K$$

The total difference in $T_s$ due to the change from 1/2" foam to 7/8" foam is 5.736°.

| 1/2" | | 7/8" | |
|------|---------|------|---------|
| | 0.032° | | 0.158° |
| | 1.56 ° | | 7.17 ° |
| | 1.592° | | 7.328° |

Bringing all the elements of $T_s$ together we get

$$T_s = T_{ant} + (L_1 - 1)T_0 + L_1 T_{LNA} + \frac{L_1(L_2 - 1)T_o}{G_{LNA}} + \frac{L_1 L_2 T_{RX}}{G_{LNA}}$$

Substituting values,

222

$$T_s = 26 + (1.023 - 1)315 + (1.023)(120) + \frac{(1.023)(50.118)(315)}{100,000}$$

$$= 26 + 7.245 + 122,76 + 0.161 + 7.17$$

$$= 163.336°K$$

## III. Calculate G/T

G/T for this system is, then,

$$G/T = G_a - 10 \log_{10} T_s$$

$$= 44.4 - 10 \log_{10} 163.336$$

$$= 44.4 - 22.1308$$

$$= 22.26 \text{ dB/°K}$$

Remember that a change in feedlines make a difference in $T_s$ of 5.736°. The difference in G/T would be

$$G/T = G_a - 10 \log_{10} T_s$$

$$= 44.4 - 10 \log_{10} (163.336 - 5.736)$$

$$= 44.4 - 21.97$$

$$= 22.42 \text{ dB/°K}$$

for an increase in G/T of only 0.164 dB/°K.

## C. Determine BW$_{RF}$

The *receiver noise bandwidth* takes into account the thermal noise produced in a receiver by the random movement of electrons. This is included in the expression for calculating C/N by the term:

$$N_{RX} = 10 \log_{10} B$$

Where: B = receiver noise bandwidth in Hz

In this system, the receiver has a noise bandwidth of 27 MHz (27,000,000 Hz). Using the expression for the receiver noise contribution:

$$N_{RX} = 10 \log_{10} B + 10 \log_{10} (2.7 \times 10^7)$$

223

$$= 10 \, (7.431)$$

$$= 74.31$$

## D. Determine EIRP

Each of the individual calculations in the expression for C/N is now complete.

The one remaining piece of information necessary is the EIRP for SATCOM I. Examining a footprint map for SATCOM I, we find that our earth station location (in Massachusetts) lies between contours of EIRP = 33 dBw and EIRP = 34 dBw. To be conservative in our calculations, we will use the worst-case value of 33 dBw.

## E. Calculate C/N

We have now calculated each of the factors in the expression for C/N.

$$C/N = EIRP - S + G/T - 10 \log_{10} B + 228.6$$

Substituting our values:

Where: EIRP = 33 dBw

$$S = 196.5 \, dB$$

$$G/T = 22.26 \, dB/°K$$

$$B = 27 \, MHz$$

$$= 2.7 \times 10^7 \, Hz$$

$$10 \log_{10} B = 74.31$$

228.6 = Boltzman's Constant

$$C/N = EIRP - S + G/T - 10 \log_{10} B + 228.6$$

$$= 33 - 196.5 + 22.26 - 74.31 + 228.6$$

$$= 13.05 \, dB$$

This is a "downlink" C/N, sometimes referred to as a "clear sky" C/N.

## F. Calculated Interference

Another important consideration in system design is interference.

224

Interference comes from basically two sources: terrestrial microwave (the 3.7 to 4.2 GHz band is shared with common carrier terrestrial links) and other satellites. A carrier-to-interference (C/I) ratio of 18 dB is just perceptable to the viewer. Total interference to a system is given by

$$C/I_{TOTAL} = C/I_{ADJ\text{-}SAT} + (C/I)_{TER} + (C/I)_{INT}$$

Where: $C/I_{ADJ\text{-}SAT}$ = adjacent satellite interference

$C/I_{TER}$ = terrestrial interference

$C/I_{INT}$ = internally generated interference in satellite

$+$ = refers to power additions

Combining downlink $C/N_d$ with $C/I$:

$$C/N_T = 10 \log_{10} \left[ \cfrac{1}{\cfrac{1}{antilog_{10}\left(\frac{C/N_d}{10}\right)} + \cfrac{1}{antilog_{10}\left(\frac{C/I}{10}\right)}} \right]$$

and using the values:

$C/N_d = 13.05$ dB

$C/I = 18$ dB (worst case situation)

we obtain

$$C/N_T = 10 \log_{10} \left[ \cfrac{1}{\cfrac{1}{antilog_{10}\left(\frac{13.05}{10}\right)} + \cfrac{1}{antilog_{10}\left(\frac{18}{10}\right)}} \right]$$

$$= 10 \log_{10} \cfrac{1}{\cfrac{1}{0.0495} + \cfrac{1}{0.0158}}$$

$$= 11.84 \text{ dB}$$

The downlink C/N calculated in Section 3e is actually lowered by the C/I contribution. In this case, the reduction is $13.05 - 11.84 = 1.21$ dB.

## 4. Calculate S/N

### A. Compute Theoretical (Downlink) S/N

Using the previously derived approximations for $V_{S/N}$ and $A_{S/N}$:

$$V_{S/N} \approx C/N + 39.8$$

$$A_{S/N} \approx C/N + 47.85$$

We obtain,

$$V_{S/N} \approx C/N + 39.8$$

$$\approx 13.05 + 39.8$$

$$\approx 52.85 \text{ dB}$$

$$A_{S/N} \approx C/N + 47.85$$

$$\approx 13.05 + 47.85$$

$$\approx 60.9 \text{ dB}$$

### B. Compute Actual (Including C/I) S/N

$$V_{S/N} \approx C/N_T + 39.8$$

$$A_{S/N} \approx C/N_T + 47.85$$

We obtain,

$$V_{S/N} \approx 51.61 \text{ dB}$$

and

$$A_{S/N} \approx 59.69 \text{ dB}$$

### C. Determine Margin

The values we have calculated for $V_{S/N}$ and $A_{S/N}$ are theoretical. That is, they were calculated under the assumption that all system components were aligned and working perfectly. We now have to consider *margin*.

Various factors can degrade the performance of an earth station. Some are beyond the user's control, some are not. A margin in C/N is necessary to compensate for miscellaneous and variable downlink losses due to atmospheric absorption, antenna mispointing, rain attenuation, system degradation (e.g., decrease in EIRP), polarization misalignment, or receiver noise temperature increase. Nominal values for these factors add up to about 2 dB. If worst-case conditions all occur simultaneously, then C/N degradation could be almost 5 dB. However, it is extremely improbable that worst-case conditions will all occur simultaneously.

If we decide that our margin should be a nominal 2 dB, then

$$V_{S/N} \approx C/N + 39.8 - \text{margin}$$

$$\approx 11.84 + 39.8 - 2$$

$$\approx 49.64 \text{ dB}$$

$$A_{S/N} \approx C/N + 47.85 - \text{margin}$$

$$\approx 11.84 + 47.85 - 2$$

$$\approx 57.69 \text{ dB}$$

If we want higher audio and video signal-to-noise ratios, we must increase C/N to compensate for the decrease in S/N caused by the margin. To increase C/N, we must increase G/T, our controllable variable in the expression for C/N, by the amount lost due to margin. Specifically, we must choose a new antenna/LNA combination to give us the desired increase in G/T.

All the parameters of system design discussed here basically tie into the G/T calculations. The system designer should be aware that it will probably take more than one calculation of G/T to determine the antenna/LNA combination necessary to meet system design objectives.

## 5. ANTENNA ALIGNMENT

In this section, we will consider *LOOK ANGLE* and *MAGNETIC DECLINATION*.

### Look Angle Calculations

The look angle of an antenna is its azimuth and elevation when it is pointed at the desired satellite. The equations for calculating azimuth (Az) and elevation (El) are:

$$Az = 180 + \cos^{-1} \frac{\tan (LAT)}{\tan Y}$$

$$El = \tan^{-1} \frac{\cos Y - 0.15166}{\sin Y}$$

Where: Az = antenna azimuth in degrees from true north.

El = antenna elevation in degrees above the horizon.

LONG = earth station longitude

LAT = earth station latitude

SAT = satellite longitude

$Y = \cos^{-1}$ [cos (LONG − SAT)] [cos (LAT)]

Degrees north and degrees west are positive.

*EXAMPLE.* To calculate the look angle for an earth station at 42° 15′ 45″ North and 71° 9′ 48″ West, and Westar I and IV at 99° West, calculate as follows:

$Y = \cos^{-1}$ [cos (LONG − SAT)][cos (LAT)]

$\quad = \cos^{-1}$ [cos (71° 9′ 48″ − 99°)][cos (42° 15′ 45″)]

$\quad = \cos^{-1}$ [0.883][0.741]

$\quad = 49.069$

$Az = 180° + \cos^{-1} \dfrac{\tan (LAT)}{\tan Y}$

$Az = 180° + \cos^{-1} \dfrac{\tan (42° 15′ 45″)}{\tan 49.069}$

$Az = 180° + \cos^{-1} \dfrac{0.9052}{1.1532}$

$\quad = 180° + 38.276$

$\quad = 218° 27′$

$El = \tan^{-1} \dfrac{\cos Y - 0.15166}{\sin Y}$

$\quad = \tan^{-1} \dfrac{\cos 49.069 - 0.15166}{\sin 49.069}$

$$= \tan^{-1} \frac{0.655 - 0.15166}{0.7555}$$

$$= \tan^{-1} \frac{0.503}{0.7555}$$

$$= \tan^{-1}(0.6664)$$

$$= 33° 67'$$

## A GRAPHICAL APPROACH TO LOOK ANGLE CALCULATIONS

The azimuth and elevation charts (Figures 13 and 14) provided below provide a graphical solution to the antenna pointing problem. The following example will explain how to use the charts.

*EXAMPLE.* To determine the look angle for an earth station at 42° 15' 45" North and 71° 9' 48" West, determine the azimuth and elevation as follows:

$$\Delta L = |LONG - SAT|$$

Where: LONG = earth station longitude

SAT = satellite longitude

$$\Delta L = |71° 9' 48" - 99°|$$

$$= |-28|$$

$$= 28°$$

To find the azimuth, locate the earth station latitude (42° 15' 45") along the horizontal axis of the azimuth chart (Figure 14). Draw a vertical line perpendicular to this axis from that point. Find the intersection of this line and the $\Delta L$ curves ($\Delta L = 28°$) closest to the calculated value of $\Delta L$. Interpolate for the calculated value of $\Delta L$. Draw a line parallel to the horizontal axis, from the $\Delta L$ point on the latitude line to the vertical axis.

Read the azimuth value off the vertical axis. As the earth station in this example is east of the satellite, read from the " − " column. The value is slightly less than 220°. Our calculated value was 218°.

To find the elevation, locate the earth station site latitude (42° 15' 45") along the horizontal axis of the elevation chart (Figure 15). Draw a vertical line perpendicular to this axis from that point. Find the inter-

'ΔL' IS THE ABSOLUTE VALUE DIFFERENCE BETWEEN THE EARTH
STATION ANTENNA SITE LONGITUDE AND THE SATELLITE LONGITUDE

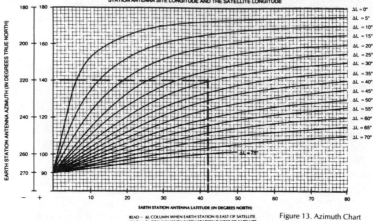

READ – ΔL COLUMN WHEN EARTH STATION IS EAST OF SATELLITE.
READ + ΔL COLUMN WHEN EARTH STATION IS WEST OF SATELLITE.

Figure 13. Azimuth Chart

'ΔL' IS THE ABSOLUTE VALUE DIFFERENCE BETWEEN THE EARTH
STATION ANTENNA SITE LONGITUDE AND THE SATELLITE LONGITUDE

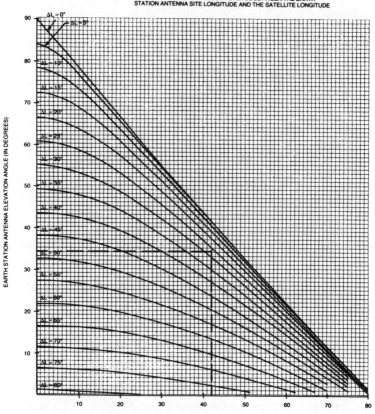

EARTH STATION ANTENNA SITE LATITUDE (IN DEGREES NORTH)    Figure 14. Elevation Chart

230

section of this line and the ΔL curves (ΔL = 28°) closest to the calculated value of ΔL. Interpolate for the calculated value of ΔL. Draw a line from the ΔL point on the latitude line, parallel to the horizontal axis, to the vertical axis.

Read the elevation value off the vertical axis. Note that the value is approximately 34°. Our calculated value was 34.1°.

## NOISE CALCULATIONS

**Series Noise Temperature.** For N devices in series, each with an effective input noise temperature $T_e(N)$, the effective input noise temperature for the system is

$$T_e(1)...T_e(N) = T_e(1) + T_e(2) + T_e(3) + ... + T_e(N)$$

**Series Noise Figures.** In similar fashion, the formula for N devices in series, each with a noise figure $NF_N$ and gain $G_N$, the system noise figure is:

$$NF_1 ... NF_N = NF_1 + \frac{NF_2 - 1}{G_1} + \frac{NF_3 - 1}{G_1 G_2}$$

$$+ ... \frac{NF_N - 1}{G_N \_ G_N} \qquad \frac{NF_N - 1}{G_1 G_2 ... G_{N-1} G_N}$$

**Converting Noise Temperature to Noise Figure.** To convert effective input noise temperature T in degrees Kelvin to noise figure in dB:

$$dB = 10 \log \frac{T}{290} + 1$$

**Converting Noise Figure to Noise Temperature.** To convert noise figure to effective input noise temperature:

$$T = \text{antilog} \frac{dB}{10} - 1 \quad 290$$

## SYSTEM CARRIER-TO-NOISE (C/N) TEST

This is a simple test to measure the direct difference between the total system noise and the total system signal (carrier).

Finding the actual amount of carrier present is straight forward. System carrier plus system noise plus system interference are measured when the antenna is pointed at the satellite. System noise, which includes sky, solar, and thermal noise, is measured when the carrier is removed by taking the antenna off the satellite and pointing the antenna at clear sky.

The procedure is as follows:

Step 1. Disconnect prime power from the receiver.

Step 2. Disable the receiver's threshold extension.

Step 3. Place the AGC/MGC Switch into the MGC position.

Step 4. Connect a field strength meter or power meter to the receiver's 70 MHz IF Monitor Port.

Step 5. Connect prime power to the receiver.

Step 6. Adjust the receiver's manual gain control for a power reading of $-20$ dBm on the power meter or 29 dBmV on the field strength meter. This insures that the IF Amplifier is in its linear gain region and not in saturation.

Step 7. Measure and record the carrier plus noise power $(C + N)$.

Step 8. Raise the antenna at least $3°$ but no more than $6°$. This takes the antenna off the satellite and leaves only output due to system noise present at the IF Monitor Port. Be sure that the antenna is not pointed at a region of the sky where the sun is located. The sun acts as a large thermal noise source and will throw off any measurements made.

Step 9. Measure and record the noise power $(N)$.

Step 10. Return the antenna to its original position.

Step 11. Calculate the system carrier-to-noise ratio by the formula:

$$C/N = 10 \log_{10} \frac{(C + N) - N}{N}$$

Where: $C$ = carrier power in mW

$N$ = noise power in mW

This value should agree closely with calculated values.

## NOTE

If a field strength meter is used, convert the voltages measured to milliwatts.

Step 12.  Remove the prime power from the receiver.

Step 13.  Disconnect the Power Meter or Field Strength Meter from the 70 MHz IF Monitor Port of the Receiver.

Step 14.  Enable the receiver's threshold extension.

Step 15.  Place the AGC/MGC switch in the AGC position.

Step 16.  Connect prime power to the receiver.

This completes the carrier-to-noise measurements.

# Glossary

**A-$\frac{2}{3}$**—Designation for Anik 2 and 3, 114 degrees W.

**A/B**—Designation for Anik B, 109 degrees W.

**absolute scale**—See Kelvin scale.

**ac**—Alternating current or voltage.

**active component**—Unit capable of amplifying a signal. A device that requires the use of an operating voltage and whose output is usually greater than its input.

**AD**—Designation for Anik D, 104 degrees W.

**af**—Audio frequency. Generally considered to be any frequency in the range from 20 Hz to 20 kHz.

**afc**—See automatic frequency control.

**aft**—Automatic fine tuning.

**agc**—See automatic gain control.

**alc**—Automatic level control. Synonymous with automatic gain control.

**AM**—See amplitude modulation.

**amplifier**—An active component. A unit that requires an external source of operating power. Usually a circuit whose output is greater than its input.

**amplitude modulation**—A method of loading a signal onto a higher frequency carrier wave. The carrier's amplitude changes in accordance with the strength of the modulating signal. Abbreviated as AM.

**Anik**—Canadian satellite system.

**Anik B**—Canadian satellite positioned at 109 degrees W.

**Anik D**—Canadian satellite positioned at 104.5 degrees W.

**Anik 3**—Canadian satellite positioned at 114 degrees W.

**aperture**—Commonly taken to mean the diameter of a dish. An indirect reference to the area of a dish. Sometimes also used to indicate the port of the dish through which the downlink signal is fed.

**apogee**—The maximum distance achieved by a satellite in its elliptical orbit following launch, prior to achieving its geocentric orbit.

**attenuation**—A decrease in signal strength of a signal, generally caused in passing through some passive component, such as coaxial cable or a video switching unit.

**audio subcarrier**—One or more carrier frequencies modulated by monophonic (mono) or stereophonic (stereo) audio signals. Satellite audio subcarriers are between 5 MHz and 8 MHz.

**automatic frequency control**—A circuit for maintaining constancy of a selected frequency or a band of frequencies. Abbreviated as afc.

**automatic gain control**—A circuit for maintaining a signal level at some predetermined amount.

**Az/El**—Azimuth/elevation.

**Az/El mount**—A support for a dish requiring two separate adjustments for the dish to enable it to *see* a selected satellite.

**azimuth**—Also known as angle of azimuth. The angular measurement in a horizontal plane, left or right of true north or south.

**balanced cable**—Coaxial cable having two independent wire conductors covered by braid used as an electrical shield. Also known as balanced coaxial cable.

**balun**—A matching transformer. A device used to match impedances of components that are to be connected. An acronym composed of balanced to unbalanced.

**bandpass filter**—Circuit designed to have a pair of cutoff points, one having a frequency lower than the other. The filter passes all frequencies between the cutoff points; attenuates all frequencies above and below.

**bandwidth**—A range of frequencies from the lowest of a selected group to the highest.

**baseband**—The group of video and audio signals, plus synchronizing and blanking pulses, including the color subcarrier, that constitute the standard NTSC signal for color TV broadcasting. Known as the video baseband, it extends from about 50 Hz to 4.2 MHz.

**bird**—Any satellite.

**blanking pulse**—A rectangularly shaped pulse used to cut off scanning action in picture tubes in TV receivers and pickup tubes in video cameras.

**block downconverter**—A frequency-changing component used for converting the frequency of the downlink band to a lower frequency for use by a number of satellite receivers.

**BNC**—A twist lock type connector used for coaxial cable.

**boresight point**—The center of the radiation pattern of a downlink signal.

**broadcast**—Generally used in reference to any terrestrial transmitted signal intended for indiscriminate reception. Examples are AM, FM and television broadcasts.

**C band**—A band of microwave frequencies used by downlink signals in the range of 3.7-4.2 GHz and by uplink signals from 5.9-6.4 GHz.

**cable television**—The distribution of television programs from satellites via coaxial cable to subscribers. An alternative to broadcast TV.

**cable TV signal converter**—An accessory device that converts cable signals to the UHF band.

**Cancom**—Canadian Satellite Communications, Inc.

**captioning**—See closed captioning.

**carrier**—Radio wave of a constant frequency that can be modulated by video and/or audio signals. A high frequency wave that can carry video or audio signals over great distances.

**Cassegrain feed**—A dish using a supplementary reflector with the signals being radiated to the center of the dish. The supplementary reflector is sometimes known as a hyperboloidal subreflector.

**CATV**—Community antenna television. Also known as cable TV. An arrangement in which homes are linked via coaxial cable to a program distributing source.

**channel**—A band of frequencies used to carry information in the form of data, telephone conversations, news, and entertainment programs such as audio and video. Channels have varying widths depending on the amount of information they must carry.

**characteristic impedance**—The impedance of coaxial cable independent of its length, and based on the physical characteristics of the cable.

**chroma**—Abbreviation for chrominance. Color signals. The hue and saturation of a color.

**chrominance signals**—Sidebands of the color subcarrier. These sidebands carry information about the hue and saturation of colors.

**circular polarization**—A type of satellite signal polarization in which the transmitted wave assumes helical form. Clockwise and counterclockwise circularly polarized signals can be transmitted simultaneously. A form of polarization used by Intelsat satellites.

**Clarke, Arthur C.**—Science fiction writer who first suggested the concept of satellite television.

**Clarke Belt**—Also known as Clarke orbit. The orbital path followed by geosynchronous satellites.

**closed captioning**—Printed copy along the base of a television picture that can only be seen on TV sets equipped with a caption decoder.

**c/n**—Carrier signal-to-noise signal power ratio. The result is expressed in dB. Sometimes written c/no and known as the carrier-to-noise power density ratio.

**coax**—Commonly used abbreviation for coaxial cable.

**coaxial cable**—Shielded cable available in two forms: balanced and unbalanced. Flexible metallic braid is commonly used as shield against unwanted signal pickup. Unbalanced cable has single central wire conductor; balanced cable has a pair of wire conductors. Characteristic impedance ranges from 50 ohms to 75 ohms. See also transmission line.

**cold lead**—The outer shield braid of unbalanced coaxial cable. The braid has a double function: it works as a signal interference shield and also as a signal carrier.

**color contamination**—This consists of bursts of color that appear on a black-and-white test pattern or picture.

**color edging**—Unwanted colors that appear along the edges of colored areas in the picture, but without any color relationship to those areas.

**color subcarrier**—A separate carrier having a frequency of 3.579545 MHz. Its function is to position the color sidebands correctly with respect to the video signal.

**common carrier**—Any organization that offers its services for transportation or communications. A trucking company is a common carrier; so is a radio station. Telephone companies and those that have satellite transponders for lease are common carriers.

**compass heading**—Magnetic direction indicated by a compass.

**component system**—See integrated component.

**composite video signal**—A signal that consists of video, audio, horizontal and vertical synchronizing (sync) pulses, and blanking pulses. The video and audio are transmitted on separate carriers; the color signals on a subcarrier.

**Comstar 2**—Satellite positioned at 95 degrees W.

**Comstar 3**—Satellite positioned at 87 degrees W.

**confetti**—Electrical noise appearing on a television screen in the form of moving dots or tiny rectangles in color.

**connectors**—Devices mounted at the ends of cables to permit easy connection and disconnection. Male connectors are plugs; female connectors are jacks.

**cps**—Abbreviation for cycles per second. Now replaced by Hertz, abbreviated as Hz. 60 cps is the same as 60 Hz.

**cross polarization**—The vertical and horizontal polarization of downlink satellite signals. Also known as opposite sense polarization.

**crt**—Abbreviation for cathode ray tube. A TV picture tube. The pickup tube used in a video camera.

**cryogenically cooled parametric amplifier**—LNA cooled close to absolute zero, equivalent to zero Kelvins.

**cutoff frequency**—The minimum operating frequency of a waveguide.

**D-2**—Designation for Comstar 2, 95 degrees W.

**D-3**—Designation for Comstar 3, 87 degrees W.

**D-4**—Designation for Comstar D-4, 127 degrees W.

**dB**—See Decibels.

**dBf**—The decibel using one femtowatt as the reference. A femtowatt is one-quadrillionth of a watt. In exponential terms it is $10^{-15}$ watt.

**dBi**—Decibels relative to isotropic. Isotropic means identical in all directions. Used to indicate the gain of a dish.

**dBm**—The decibel using one milliwatt (a thousandth of a watt) as the reference.

**DBS**—Direct Broadcasting Satellite services. Transmissions in the K band, 11.7 to 12.7 GHz.

**dBw**—The decibel using one watt as the reference.

**dc**—Direct current or voltage.

**decibels**—Abbreviated as dB. A measurement between a pair of voltages, currents or powers. Also used to compare voltage, current, or power against a reference. A logarithmic ratio.

**decoder**—Device for unscrambling video signals that have been coded.

**dedithering**—The method used to overcome the dithering process used prior to the transmission of a signal. See dithering.

**deemphasis**—A process of deliberately weakening certain frequencies that have been preemphasized prior to transmission.

**definition**—The amount of fine detail visible in a television picture.

**demodulate**—The process used to recover modulated signals from a carrier wave.

**descrambler**—See decoder. A circuit or device for unscrambling a signal and restoring it to its original NTSC format.

**direct broadcasting satellite services**—See DBS.

**dish**—Sometimes referred to as an antenna. A reflector used for collecting and focusing microwave energy.

**dish illumination**—The area of a dish *seen* by the feed.

**dithering**—The process of dispersing the video signal at a low frequency rate, generally 30 Hz.

**DNR**—Dynamic Noise Reduction. A circuit for the removal of hiss and other unwanted electrical noise.

**Domsat**—Acronym for domestic satellite transmissions.

**double reflector system**—See Cassegrain feed.

**downconvert**—The reduction of a high frequency to a lower. It usually involves the formation of an intermediate frequency (i-f) from a radio-frequency signal.

**downconverter**—A component for reducing the frequency of downlink signals.

**downlink**—Satellite to earth transmissions between 3.7 GHz and 4.2 GHz.

**dual feed**—A feed arrangement capable of utilizing both vertically and horizontally polarized signals.

**dual orthomode coupler**—Device for delivering both vertically and horizontally polarized signals to a dual LNA.

**duplex**—Two-way simultaneous communications using different frequencies.

**earth station**—Any terrestrial station used for the transmission and reception of satellite signals.

**Eb/No**—A comparison of the energy per bit to noise power density.

**EIRP**—Effective isotropic radiated power. The strength or energy level of a beamed signal, expressed in dBw.

**electrolysis**—A type of corrosion that takes place when dissimilar metals are joined.

**electron beam**—Electrons produced by a gun in a picture tube that are then focused and moved forward by a high voltage.

**electronically cooled paramp**—See uncooled parametric amplifier.

**encoding**—Signal scrambling.

**eng**—Electronic news gathering.

**equatorial orbit**—The orbit followed by satellites in the equatorial plane.

**F-1-2**—Designation for Satcom 1-2, 119 degrees W.

**F-3**—Designation for Satcom 3R, 131 degrees W.

**F-4**—Designation for Satcom 4, 83 degrees W.

**F-5**—Designation for Satcom 5, 143 degrees W.

**F connector**—Small, metallic male connector, threaded internally. Connected to coaxial cable to permit joining to female F-fitting. Baluns are often equipped with a female F-connector for 75-ohm coax and spade lugs on the other end for connection to 300-ohm antenna input terminals.

**F plug**—A male connector used on coaxial cable for joining video components.

**FCC**—Federal Communications Commission.

**feed**—See feedhorn.

**feedhorn**—The first section of the waveguide receiving the downlink signals reflected from the dish. The feedhorn is mounted at the focal point.

**fiberglass**—A plastic coating put on a dish at the time it is manufactured as a protection against the weather.

**field of view**—The unobstructed arc of sky that can be *seen* by a dish.

**figure of merit**—A quality factor characterizing the ratio of dish gain to system noise temperature.

**FM**—See frequency modulation.

**footprint**—Contour lines of EIRP indicating signal strength of downlink signals in geographic areas. The signal strength pattern of a satellite's transmission. The contour lines, drawn on a map, indicating regions of equal signal strength.

**frequency**—Number of complete waves per second. Measured in cycles per second, designated as cps or Hertz. Hertz is now the more commonly accepted form.

**frequency coordination**—A test procedure used to determine the possible level of interference from other microwave sources at specific locations.

**frequency modulation**—Abbreviated as FM, it is a method of loading a signal, audio or video, onto a higher frequency carrier. The frequency of the carrier changes in accordance with the amplitude of the modulating voltage.

The loudness of the applied audio is indicated by the amount of deviation of the carrier frequency. The frequency of the modulating signal is represented by the number of times the carrier frequency changes per second. Frequency modulation is less subject to electrical interference than amplitude modulation.

**frequency response**—The ability of a component to amplify and/or pass a band of frequencies without altering them.

**frequency reuse**—The ability of a transponder to transmit two channels simultaneously on the same frequency through the use of alternate polarization. One modulated carrier is vertically polarized; the other horizontally polarized.

**G-1**—Designation for Galaxy 1, 134 degrees W.

**gain**—Amplification of an audio or video signal. The ratio of the signal output of a component compared to the input. A comparison of an antenna with a standard. Measured in decibels.

**GCE**—Ground Communications Equipment. Electronics components used in an earth station.

**geodetic north**—True north.

**geodetic south**—True south.

**geostationary**—See geosynchronous orbit.

**geosynchronous orbit**—The matching of an orbital day with an earth day. Under these conditions, the satellite hovers in a fairly fixed position relative to a small earth area.

**gigahertz**—1,000 megaHertz or 1,000 million cycles. Abbreviated as GHz.

**global beam**—Broad pattern radiation of a satellite signal.

**global beam antenna**—A wide beam antenna used on a satellite so signal covers more than a third of the earth's surface.

**g/t**—Ratio of dish gain compared to noise temperature. Expressed in dB.

**guided wave**—A radio wave traveling through a conductor.

**hardline**—A type of coaxial cable using a solid metal shield instead of flexible shield braid.

**heterodyning**—The beating or mixing of two signals, usually an incoming signal and a signal produced by a local oscillator. The result is a new frequency known as the intermediate frequency, abbreviated as i-f.

**high pass filter**—A circuit designed to pass all frequencies above a selected cutoff point and to attenuate all frequencies below it. High pass filters are used at the antenna terminals of a TV receiver to remove interference that may be caused by *ham* and CB radio transmissions.

**high Z**—High impedance.

**horn**—Flared type of waveguide, sometimes referred to as a horn antenna.

**hot lead**—The center conductor of unbalanced coaxial cable.

**hydrazine**—Fuel used by satellites for their thruster rockets.

**hyperboloidal subreflector**—See Cassegrain feed.

**Hz**—Abbreviation for Hertz or cycles per second (cps).

**IC**—Integrated circuit.

**i-f**—Intermediate frequency. The frequency following the downconversion process, generally 70 MHz. A frequency produced by the heterodyning of an incoming signal and a local oscillator signal. The i-f is always lower in frequency than the two frequencies producing it.

**illumination**—See dish illumination.

**IM**—See intermodulation distortion.

**image generation**—The production of image frequencies by a mixer. Identical intermediate frequencies produced by a pair of dissimilar input signals.

**impedance**—Total opposition to the flow of a varying current. The vector sum of resistance and reactance. Measured in ohms. Represented by the letter Z.

**integrated component**—A unit containing two or more independent sections on a single chassis. These sections are sometimes packaged individually to form a component system.

**Intelsat**—International Television Satellite Organization.

**intermodulation distortion**—Abbreviated as IM, it is the production of spurious responses that appear at the output of a mixer circuit.

**ionized layers**—Rings of charged particles ranging from 60 to 200 miles above the earth's surface.

**isolator**—Device that permits the passage of signals in one direction while attenuating them in the other.

**Jack**—A female connector. Used on cables and components. See connectors.

**K band**—Band of frequencies from 11.7 to 12.2 GHz. Reserved for direct broadcasting satellite services. Also known as the Ku band.

**Kelvin scale**—A scale used for the measurement of noise temperatures and for the light levels. Sometimes known as the absolute scale. Readings on this scale are referred to as Kelvins (formerly degrees Kelvin). A measurement used to indicate the amount of thermal noise produced by a component such as an LNA.

**kHz**—Abbreviation for kiloHertz or 1,000 cps.

**kilohm**—A thousand ohms. See Resistance.

**kv**—Kilovolt or a thousand volts.

**latitude**—Also known as lines of latitude. Imaginary lines that encircle the earth in an east-west direction. The reference is the equator, having latitude 0 degrees. The true poles, North and South, have latitudes of 90 degrees.

**LCD**—Liquid Crystal Display.

**LED**—Light Emitting Diode. Can be used as a digital or alphanumeric indicator.

**level**—Signal strength of an audio or video signal, usually expressed in terms of volts or fractions of a volt.

**line amplifier**—An amplifier inserted in the line at any point following the downconverter. Its function is to overcome signal losses introduced by coaxial cable.

**lines of latitude**—See latitude.

**LNA**—See low noise amplifier.

**local oscillator**—An electronic generator in a receiver that produces a high frequency signal. When mixed with the received signal the result is an intermediate frequency, abbreviated as i-f. Local oscillators are used in superheterodyne receivers or downconverters.

**longitude**—Meridian lines extending from the true North Pole to the true South Pole.

**look angle**—Also known as the Position Angle. The angle at which a mount must be set to *see* a satellite.

**low noise amplifier**—Abbreviated as LNA. A high gain, solid state amplifier using GaAsfet and bipolar transistors. The gain of an LNA is related to its temperature in Kelvins. See also uncooled parmetric amplifier and cryogenically cooled parametric amplifier.

**low-pass filter**—A circuit designed to pass all frequencies up to a selected cutoff frequency, attenuating all frequencies above it.

**low Z**—Low impedance.

**luminance reversal**—Reversal of dark and light portions of an image, supplying the effect of a film negative.

**luminance signal**—The black and white portion of a color TV signal.

**luminosity**—Degree of brightness of a color, measured in brightness units or lumens.

**μA**—Microampere. A millionth of an ampere.

**mA**—Milliampere. A thousandth of an ampere.

**magnetic north**—Also known as the North Pole. The region on earth to which the north seeking end of a compass needle points.

**masking**—The ability of a strong signal to cover a much weaker signal, preventing it from being recognized.

**matching transformer**—See Balun.

**MATV**—Master Antenna Television. A system used to distribute signals from a single antenna to a large group of TV receivers. Commonly used in hotels, motels, and high rise apartment buildings.

**MDS**—An abbreviation for Multipoint Distribution System. A form of broadcast pay TV, not too well known. It uses just a single channel for sending its video programs to subscribers, using microwaves to do so. MDS sometimes gets its programs from satellite downlink signals. Subscribers require special antennas (not dishes) to pick up the signals.

**megohm**—A million ohms. See resistance.

**Meridian**—See longitude.

**MHz**—MegaHertz. One million cps.

**micron**—0.001 millimeter or 0.000001 meter.

**modulation**—A process of loading audio and/or video signals on a much higher frequency carrier wave. Modulation can be achieved by varying the amplitude, frequency, or phase of a carrier wave. Uplink and downlink signals are frequency modulated.

**modulator**—A circuit at the output of a satellite receiver or a separate component used to modulate a channel 3 or a channel 4 carrier wave for delivery of the receiver's baseband signals to the antenna terminals of a TV receiver. Sometimes referred to as an rf adapter or remodulator.

**mono**—Monophonic or single channel sound.

**monochrome**—Black and white.

**Mouse**—Minimum orbital unmanned satellite of the Earth, launched in 1956.

**mpx**—See multiplex.

**MSATV**—Master Satellite Antenna Television.

**MTV**—Music Television.

**Multiple LNA's**—A pair of LNA's, one for vertically polarized signals, the other for horizontally polarized signals.

**multiplex**—An electronic method for combining a pair of independent signals.

**narrow beam antenna**—High gain, directional antennas used on some satellites.

**noise**—Any unwanted electrical signal. There are several types of noise, but all are amplitude modulated signals.

**noise factor**—S/N at the output compared to S/N at the input.

**noise figure**—Total amount of noise at the output of a component compared to the noise at the input.

**noise temperature**—The noise of an LNA in Kelvins.

**noncryogenically cooled preamp**—See uncooled parametric Amplifier.

**opposite sense polarization**—See cross polarization.

**orbit inclination**—See tilt angle.

**orbital day**—The time of rotation of a satellite. Equivalent to an earth day.

**orbital slot**—Specific position of a satellite in space.

**orthomode coupler**—See dual orthomode coupler.

**Palapa B**—An Indonesian satellite. Palapa is an Indonesian expression meaning *fruit of effort*.

**parabola**—A curve produced by sending a cutting plane through a right cone parallel to a slant edge.

**parabolic dish**—A satellite dish whose reflecting surface has a parabolic shape.

**perigee**—The point nearest the earth in the orbit of a newly launched satellite.

**petallized dish**—A dish whose reflecting surface is made up of a number of individual sections called petals.

**photovoltaic cell**—Transducer for converting solar energy to electrical energy.

**plug**—See connectors.

**polar mount**—A dish support that is a celestial tracking device.

**position angle**—See look angle.

**prime focus feed**—An arrangement in which the feed is positioned at the focal point of the dish.

**probe**—The actual antenna used for receiving the satellite signal. The output of the probe is brought into the LNA.

**quasi DBS**—A form of satellite transmission using the C band and having higher power transponders. Dish size is about 4 feet.

**quasi-optical**—Waves in the C and K bands. These have some of the properties of light waves.

**reception window**—Rectangular area in front of the dish whose imaginary outline defines the limits of movement of the dish to receive signals of a selected satellite.

**remodulation**—The process of demodulating the baseband signals and then modulating them again, using a different modulation process.

**resistance**—Opposition to the flow of current, whether ac or dc. Expressed in ohms (or multiples of the ohm), the basic unit of resistance.

**rf adapter**—See modulator.

**RG-58U**—Coaxial cable having a characteristic impedance of 50 ohms.

**RG-59U**—Coaxial cable having a characteristic impedance of 75 ohms.

**Satcom 1**—Satellite positioned at 135 degrees W.

**Satcom 3R**—Satellite positioned at 131 degrees W.

**Satcom 4**—Satellite positioned at 83 degrees W.

**satellite**—Any object in space, manmade or natural, orbiting the earth. Satellites are either geostationary, or have rotational speeds faster or slower than the earth.

**satellite bandwidth**—The bandwidth for transponders operating in the C band is 40 MHz. The total bandwidth for the C band is 500 MHz.

**satellite receiver**—A component that selects a particular satellite for signal reception. The receiver may include a downconverter (but more often the downconverter is a separate unit) and a demodulator. Output of the receiver is a remodulated signal using the carrier frequency of either channel 3 or channel 4.

**scalar feed**—A number of concentric rings placed at the entrance to the feedhorn. See feedhorn.

**SCPC**—Single Channel per Carrier. A technique for the transmission of audio signals by a transponder.

**servo**—Abbreviation for servomechanism. A motor system whose angular rotation can be remotely controlled and measured.

**shroud**—A metallic ring fastened to the perimeter of a dish. The width of the shroud ranges from 6″ to 18″ and is used to prevent alternate channel sidelobe pickup and electrical interference.

**side lobe**—A minor signal lobe radiated by an antenna.

**signal combiner**—Device that permits the simultaneous transmission of more than one uplink signal.

**signal strength meter**—A ballistic type of meter that supplies an indication of the strength of a received signal.

**signal switcher**—A component for switching program sources to a TV set. Permits the use of a number of sources using the same TV receiver but without the need for connecting and disconnecting cables.

**single feed**—One capable of receiving satellite signals polarized in one direction only.

**site tester**—Device used to determine the possible locations of a dish.

**skin effect**—The tendency of a varying current to travel on the surface of a conductor. A characteristic noted with high frequency currents.

**slot**—Space occupied by a satellite. The longitudinal position of a satellite.

**s/n**—Signal-to-noise ratio. A comparison between the amount of signal voltage compared to the amount of noise voltage.

**snow**—Electrical interference in a monochrome picture. Consists of black and white dots in motion, superimposed on the received picture.

**sparklies**—Tear shaped drops appearing horizontally in a weak picture received via satellite.

**specs**—Abbreviation for specifications. A listing of the mechanical and electronic characteristics of a component.

**spherical dish**—A dish whose surface follows the contour lines of a sphere.

**Sputnik**—First manmade satellite. Launched into space by the USSR in 1957.

**station keeping**—Adjustments to the orbit of a satellite. Controlled by earth located telecommunications equipment.

**statute mile**—A mile having a length of 5,280 feet. A standard mile.

**stereo**—Stereophonic or two-channel sound.

**Stereo Processor**—A component for the processing of mono and stereo downlink audio signals.

**subcarrier**—A supplementary carrier separated from the main carrier by a fixed amount in terms of frequency. Modulation of the subcarrier may be the same as that used for the main carrier, or different.

**switcher**—Passive device for switching a variety of program sources to the TV set.

**Syncom**—First geosynchronous North American satellite, launched in 1963.

**teleconference**—A *meeting* in which a conference group makes use of an uplink signal and a number of downlinks, plus telephone communications.

**telemetry**—The transmission of data by wire, radio, or signals transmitted to and from satellites for recording or analysis. Measurements made at a distance.

**Telstar**—First satellite to relay a TV picture. Launched in 1962.

**thermal noise**—Electrical noise produced by the heat agitation of molecules near the surface of the earth.

**threshold**—Level of carrier to noise.

**tilt angle**—Angular shift of a satellite with respect to the equatorial plane. Also known as orbital inclination.

**translation frequency**—The frequency difference between an uplink and downlink signal. It is 2,220 MHz.

**transmission line**—Conductors used for guiding waves. A connecting link between an antenna and a receiver. Twin lead (also known as two-wire transmission line), coaxial cable, and waveguides are types of transmission lines. An antenna lead-in.

**transponder**—Component on a satellite that receives uplink signals, converts them to downlink signals, and transmits them to earth.

**true north**—The northern axis of the earth. The axis of the earth that points to the North Star.

**TVRO**—Television receive only. A phrase applied to a station intended only for the reception of signals from satellites.

**twin-lead**—Two-wire transmission line. Impedance is usually 300 ohms.

**unbalanced cable**—Coaxial cable consisting of a central conductor and an outer conductor made of flexible shield braid. The two conductors are separated by an insulating material, often plastic.

**uncooled parametric amplifier**—An LNA having a very low noise level. More expensive than typical LNAs. Used when signal input levels are extremely low. Also known as electronically cooled paramp and non-cryogenically cooled preamp.

**unguided wave**—A wave radiated from an antenna. Guided waves are those which travel through conductors.

**unipolar transistor**—A GaAsFet transistor.

**uplink power**—The amount of transmitting power used by an earth station for the uplink signal.

**uplink signals**—Signals transmitted from earth to satellites in the frequency range of 5.9 to 6.4 gigahertz.

**video inversion**—One type of signal scrambling in which the baseband signal is changed in polarity.

**voltage standing wave ratio**—Abbreviated as VSWR. A reference to the amount of signal energy to the amount not used. Ideal VSWR is 1:1.

**VTO**—Voltage tuned oscillator. An oscillator whose frequency can be altered by changes in the applied dc voltage.

**W-2**—Designation for Westar 2, 79 degrees W.

**W-3**—Designation for Westar 3, 91 degrees W.

**W-4**—Designation for Westar 4, 99 degrees W.

**W-5**—Designation for Westar 5, 123 degrees W.

**waveguide**—A completely enclosed circular or rectangular type of metallic tubing acting as a transmission line between the scalar feed and the input to the LNA. The antenna probe is positioned at the end nearest the LNA. The waveguide is its feed.

**wavelength**—The distance from the start of a single wave to the end of that same single wave. Wavelength is frequently represented by the Greek letter lambda ($\lambda$).

**Westar 2**—Satellite positioned at 79 degrees W.

**Westar 3**—Satellite positioned at 91 degrees W.

**Westar 4**—Satellite positioned at 99 degrees W.

**Westar 5**—Satellite positioned at 123 degrees W.

**wind load survival**—Ability of a dish to withstand wind pressure.

**window**—See Reception Window.

**Z**—The letter used to represent impedance.

**zero center tuning meter**—A ballistic meter whose pointer can swing to the left or right of zero in the center of a scale. Used as an aid in tuning accuracy.

# Index